Simplified Concrete Masonry
Planning and Building

Simplified Concrete Masonry Planning and Building

J. RALPH DALZELL

Second Edition

REVISED BY

FREDERICK S. MERRITT

Consulting Engineer

McGRAW-HILL BOOK COMPANY

New York   St. Louis   San Francisco   Düsseldorf   Johannesburg
Kuala Lumpur   London   Mexico   Montreal   New Delhi
Panama   Rio de Janeiro   Singapore   Sydney   Toronto

Library of Congress Cataloging in Publication Data

Dalzell, James Ralph, 1900–1970
Simplified concrete masonry planning and building.

Published in 1955 under title: Simplified masonry planning and building.
1. Concrete construction.   2. Masonry.   I. Merritt, Frederick S.   II. Title.
TA682.4.D35   1972   693.5   72-7371
ISBN 0-07-015223-3

234567890 KPKP 76543

*The editors for this book were William G. Salo, Jr., and Lydia Maiorca,
and its production was supervised by George E. Oechsner.
It was set in Intertype Baskerville by The Maple Press Company.
It was printed and bound by The Kingsport Press.*

Contents

HOUSE PLANS—PLATES I TO VI—*follow the Index.*

Preface to Second Edition

Continuing demand for this book has made publication of a second edition necessary. In producing this edition, the publishers deemed it advisable to update the book.

The author, the late J. Ralph Dalzell, had shown great foresight in the preparation of the first edition. Nearly all the methods of concrete construction that he recommended and explained are still widely used and recommended by authorities in the field. The changes that have occurred in the many years since the first edition was published have been mainly in specifications for materials—portland cements and masonry cements. Also, in some cases, greater use is being made of newer materials, including precast concrete, prestressed concrete, plywood, and plastics. To date, these have had only a minor effect on construction methods discussed in this book. Nevertheless, the changes that have occurred are reflected in the new edition.

Frederick S. Merritt

Preface to First Edition

This book introduces an interesting, refreshing, and successful departure from the more or less precise styles employed in all previous books on masonry work.

Previous books present their subject matter and explanations in what can now be called a stiff and far too general manner. Readers find that even studied perusal of such books produces only an indistinct and difficult-to-visualize picture of masonry practices. Specific applications of materials and manipulative procedures are difficult to interpret and the books do not serve the purpose of young mechanics.

This book, on the other hand, was planned especially for use by young mechanics. The language is easy to read, easy to understand, and easy to remember. All subject matter and explanations are set forth in such a manner that they match the requirements of young mechanics and easily create the necessary visualization. Specific applications of materials and manipulative procedures can be selected to suit the specifications of any and all projects young mechanics are likely to be called on to do.

Teachers will quickly find that this book anticipates all learning difficulties and that it explains not only the *what* and *why* of masonry planning and construction, but especially the *how*. Young mechanics will find it a great help to them in all aspects of their advancement to journeyman status and even more help as they seek and carry on typical masonry work.

One of the many special features of the book is that piece-by-piece explanations and illustrations are employed to show how masonry is planned and placed or laid up. For example, all concrete-block projects are explained and illustrated in terms of each and every block to be laid in walls and other details. This procedure simplifies the learning process and helps to create better workmanship.

To a great extent, pictorial illustrations are used in connection with the text material. Such illustrations can be visualized with ease and show each masonry component in its proper relation to allied parts. Conventional section views and other representations are employed frequently enough to develop proper visualization of details from standard working drawings.

Many young mechanics may have previously used materials and procedures which by every calculation or good intent should have produced sound jobs of masonry construction. Yet, such jobs have often been faulty as to type of materials, mixture of materials, or workmanship. Unfortunately, not many mechanics have the facilities or the experience to examine and judge materials and procedures. Therefore, this book presents a general knowledge of *tested* materials and procedures which can be depended upon to produce sound masonry work.

The author heartily agrees that the use of standard tools, such as surveying instruments, should be employed wherever possible. However, there are many cases where young mechanics do not have such tools or access to them. Therefore, this book suggests the use of substitute methods of leveling and squaring which, if carefully followed, produce satisfactory results.

Other features of this book are set forth and explained in Chapter One.

The author wishes to express his appreciation to the following sources for splendid cooperation.

Portland Cement Association
Chicago, Illinois
Samuel Paul, Architect
Jamaica, New York
Walter T. Anicka, Architect
Ann Arbor, Michigan
Campbell and Wong, Architects
San Francisco, California
R. Randolph Karch
State Department of Education
Harrisburg, Pennsylvania
J. Douglas Wilson, Secretary
Chamber of Commerce
South Pasadena, California
Joseph Santucci, Masonry Contractor
Redondo Beach, California

Walter T. Reid, Masonry Contractor
    Portland, Oregon
William H. Hart, Masonry Contractor
    Des Moines, Iowa
Lester P. Sandberg, Masonry Contractor
    Chicago, Illinois
Carl E. Anderson, Masonry Contractor
    Minneapolis, Minnesota
Besser Manufacturing Company
    Alpena, Michigan
Lone Star Cement Corporation
    Chicago, Illinois
Marquette Cement Manufacturing Company
    Chicago, Illinois
Ernest G. Schau, Technical Editor
    Practical Builder Magazine
    Chicago, Illinois
National Bureau of Standards
    Washington, D.C.
Boston Society of Architects
    Boston, Massachusetts

Simplified Concrete Masonry
Planning and Building

CHAPTER ONE

Introduction

Masonry work reflects a long and honored tradition of progressive attitudes and pride of accomplishment. It is one of the oldest trades, and it has always attracted skilled workers who find enjoyment, contentment, and worthwhile careers in advancing the trade and contributing to the constantly improving quality of structural work.

The fact that masonry has survived through the ages and that it continues to move forward with the times constitutes ample proof that any young man may aspire to membership in the trade with full assurance that his life will be spent in an honorable and always rewarding manner.

The aim of this book is to provide a thorough and easily understandable introduction to and explanation of all the masonry projects which young mechanics are likely to encounter. All through the book, the explanations and illustrations, together with examples and practical applications, are set forth in terms of the trade for the guidance of inexperienced masons. Each explanation is complete. Points which usually prove difficult have been particularly stressed. The whole presentation has been organized to provide a ready reference. Planning and manipulative procedures and methods are set forth in order of importance, to supply the information and help young mechanics need.

The following general aspects of the book should be studied and understood:

Fundamentals. First, the book deals with *fundamentals.* The many explanations and illustrations were developed to provide a broad understanding of the trade and the basic principles that young mechanics have to learn before they can become proficient masons. In this sense, the book is a primer which should prove invaluable.

Planning. The book continually stresses the great importance of carefully planning all masonry projects *before* actual construction is under way. Proper planning avoids mistakes, saves materials and time, and assures the best kind of workmanship.

Design. Design is more properly a function of architects and engineers. However, the book contains a great deal of instruction on design which masons can use to advantage if they are called upon actually to do such work on jobs where architects and engineers are not available.

Manipulative Procedures. The book places the greatest stress on *how to do* all commonly encountered masonry work. The procedures given are those which are acceptable in terms of good workmanship and are designed to comply with the requirements of all building codes.

Accuracy. All masonry work requires extreme accuracy. In fact, accuracy is the watchword of the trade. With this in mind, the book includes every possible explanation in connection with level and alignment.

Support. Most masonry work acts as support for other structural details but must, itself, be amply and properly supported. The principles involved are carefully explained wherever necessary.

Building Codes. The great importance of local and national building codes and the necessity for abiding by them are pointed out.

Regional Requirements. Various regions of the country have somewhat different requirements and practices. The book explains such differences and supplies information for proper planning and construction in all regions. The book is usable by young mechanics in any part of the country.

Modular Planning. Briefly, modular planning consists of creating dimensions which are multiples of 4 and 8 inches. In other words, architects make all house and other building plans so that the principal dimensions are exactly divisible by either 4 or 8 inches. Thus, a wall-length dimension of 20′ 0″ is a modular dimension, whereas one of 20′ 3″ is not.

The benefits of modular dimensions include both material and labor economy, as well as more attractive work, especially so far as masonry is concerned.

In keeping with the great benefits of modular dimensions, material manufacturers now produce products whose sizes coincide with modular dimensions. For example, instead of the *old* concrete blocks which were 8 inches thick, 7¾ inches high, and 15¾ inches long, the newer and better modular blocks are 7⅝ inches thick, 7⅝ inches high, and 15⅝ inches long. The modular blocks, with ⅜-inch-thick horizontal and vertical joints, occupy spaces which are exactly 8 inches high and 16 inches long. Thus, they are modular in all directions, and an exact number of them will fit into any modular dimension without cutting. For example, if a wall is made 18′ 8″ (224 in.) long and 10′ 0″ (120 in.) high, it will allow the use of exactly 14 blocks for the length of the wall and will be exactly 15 blocks high.

In order to be entirely up-to-date and most useful to young mechanics, the book explains modular planning and employs it for all examples.

Reinforcement. Until recently, the use of steel reinforcement in masonry work, especially as regards concrete-block construction, was confined to a relatively small section of the country, where the occurrence of exceptional and unusual natural forces required its use as a means of avoiding cracks and structural failures.

Experiments and engineering studies in such regions of the country proved that reinforced-concrete block could be most satisfactorily and economically used for *all* types of structures. When properly reinforced, concrete-block construction has the ability to resist all unusual forces.

Somewhat later experiments and continued engineering study further proved that reinforcement is of great value in all regions of the country where severe windstorms, alternate freezing and thawing, and expansion and contraction are likely to occur. Proper reinforcement prevents cracks and many other undesirable conditions which had previously limited the use of concrete block.

The results of such experiments and study have shown, without doubt, that reinforcement is desirable in concrete-block structures located in *any* part of the country. Such reinforced-concrete-block structures have many of the advantages of reinforced concrete but at much less cost. They are rigid and durable structures which are capable of successfully resisting unusual forces and shocks of all kinds. Actually, reinforced-concrete-block structures are tied together by the combined strengths of concrete and steel. Thus, without too much extra cost, structures can be given added strength to make them a most desirable type of construction.

With the foregoing considerations in mind, the book treats concrete-block construction with and without reinforcement, but with emphasis on the importance and desirability of reinforcement.

Examples. Wherever important aspects of planning and construction occur, the book presents examples which correspond with actual job experience as closely as practical. In this way, proper visualization of the problems is stressed.

Illustrative Example. One complete chapter is devoted to the planning and construction aspects of an actual building in which masonry constitutes the principal material. The building is "planned" and "built" completely as a means of illustrating many important fundamentals and manipulative procedures.

CHAPTER TWO

Cement

From the time man first started to build, he has looked for a material that would bind stones and other natural substances into solid, shaped masses. The ancient Assyrians and Babylonians used clay for this purpose, and the Egyptians discovered a lime and gypsum mortar which they used for building the pyramids. The Greeks added further improvements, and the Romans finally developed a cement which they used in structures of great durability.

The foundations for the buildings in the Forum of Rome were made of a form of concrete. The great Roman baths, built as early as 27 B.C., and the Colosseum are examples of structural work where cement mortar was used. The Romans mixed lime with volcanic ash from Mt. Vesuvius to produce a cement, known as *pozzolana,* which was capable of hardening under water.

During the long period known in history as the Dark Ages, the art of making cement was lost and was not rediscovered until the eighteenth century. From then on, continuous progress has been made.

Early in the 1800's, an English bricklayer named Aspdin made a new cement which he called *portland* cement because it had the same general color as the stone quarried on the Isle of Portland, off the British coast. Aspdin's contribution constituted a method of proportioning limestone and clay, pulverizing them, and burning the mixture into a clinker which was ground into the finished cement.

Before portland cement was discovered, large quantities of natural cement were used. Such cement was produced by burning a natural mixture

5

of lime and clay. Because the ingredients of natural cement were mixed by nature, its properties varied, and it could not be depended upon so far as strength was concerned. Thus natural cement gave way to portland cement, which is predictable and a known product.

The manufacture of portland cement in the United States dates back to the 1870's. Probably the first plant to start production was at Coplay, Pa. Somewhat later, a firm at South Bend, Ind., also started producing a form of portland cement which was made by burning it in a piece of sewer pipe and grinding the resulting clinker in a coffee mill.

In 1902, Thomas Edison introduced the long rotary kilns of the type now used in the cement industry. From that date on, the production of portland cement has increased constantly.

The purpose of this chapter is to acquaint readers with various kinds of cement, give a brief explanation about manufacturing methods, describe the kinds of cement commonly used, and provide other important information which readers may require.

TERMINOLOGY

Cement is known as a *cementitious* material. Or, in other words, as far as concrete is concerned, cement is used to bind the concrete ingredients together to form a dense and rocklike mass. Thus, cement is actually one ingredient of concrete. It is therefore incorrect to speak of cement footings, cement block, or cement foundations. Such structural items should be referred to as *concrete* footings, *concrete* block, and *concrete* foundations.

When mortar is made using cement as the only cementitious material, the mortar can be correctly referred to as *cement* mortar.

MANUFACTURE OF PORTLAND CEMENT

When Aspdin made cement by burning powdered limestone and clay in his kitchen stove, he laid the foundation for the portland industry of today, which every year processes literally mountains of limestone, clay, cement rock, and other materials into a powder so fine it will pass through a sieve capable of holding water.

Portland cement is a carefully controlled combination of ingredients, such as lime, silica, alumina, iron oxide, and small amounts of other materials, to which gypsum is added to regulate the setting time of the resultant cement.

The exacting nature of cement manufacture requires about 80 separate operations, the use of a great deal of heavy machinery and equipment, and large amounts of electricity. Each step in the manufacturing process is checked frequently.

Two different processes are used in the manufacture of portland cement. One is the dry process, and the other is the wet. When rock is used as the principal raw material, the first step after quarrying, in both processes, is the primary crushing. The rock is fed through crushers capable of handling pieces as large as an oil drum. The first crushing reduces the rock to a top size of about 6 inches. The rock then goes to secondary crushers, or hammer mills, for reduction to approximately 2-inch size or smaller.

In the wet process, the raw materials, properly proportioned, are then ground with water, intimately mixed, and fed in the form of a slurry (containing enough water to make it of a fluid consistency) into a kiln. In the dry process, raw materials are ground, mixed, and fed to the kiln in a dry state. Otherwise, the two processes are essentially the same.

The raw material is raised to a temperature of approximately 2700°F in huge cylindrical steel rotary kilns lined with special firebrick. Kilns are frequently as much as 12 feet in diameter—large enough to accommodate an automobile and longer, in many instances, than the height of a 40-story building. Kilns are mounted with the axis inclined slightly from the horizontal. The finely ground raw material or the slurry is fed into the higher end. At the lower end is a roaring blast of flame, produced by precisely controlled burning of powdered coal, oil, or gas under forced draft.

As the material moves through the kiln, certain combinations of elements are driven off in the form of gases. The remaining elements unite to form a new substance with its own physical and chemical characteristics. The new substance, called *clinker,* is formed in pieces about the size of marbles.

Clinker is discharged red hot from the lower end of the kiln and is generally cooled in various types of mechanical coolers. The heated air from the coolers is returned to the kilns, a process which saves fuel and increases burning efficiency.

The clinker may be stockpiled for further use, or it may be conveyed immediately to a series of grinding machines. Here gypsum is added, and the process is completed. The final grinding operation reduces the clinker to a finely ground powder called *portland cement.*

Strong paper bags, which have been sealed before receiving the cement, are then filled through a small opening, or "valve," by a packing machine that automatically cuts off the flow of cement when 94 pounds have entered the bag. The 94-pound bag now in general use contains 1 cubic foot of portland cement.

TYPES OF CEMENT

Several types of standard portland cements, all of which conform to the specifications of the American Society for Testing Materials, are available for specific needs:

Type I—For use in general concrete construction when the special properties specified for any other type are not required

Type IA—Same uses as Type I but where air entrainment is desired

Type II—For use in general concrete construction exposed to moderate sulfate action or where only *moderate* heat of hydration (heat created as concrete starts to harden) can be tolerated

Type IIA—Same uses as Type II but where air entrainment is desired

Type III—For use where high early strength is required

Type IIIA—Same uses as Type III but where air entrainment is desired

Type IV—For use where only *low* heat of hydration can be tolerated

Type V—For use where high sulfate resistance is required

For most ordinary masonry work, such as footings, foundations, sidewalks, and mortar, Type I serves the purpose excellently and is the type dealers deliver unless advised to the contrary at the time the cement is ordered. In cases where faster hardening and earlier strength are required, Type III can be used to advantage. Types II, IV, and V are seldom required in commonly encountered masonry work and are not carried in stock by material dealers. In cases where such types of cement are required, they must be ordered well in advance of the time they will be needed.

White Cement. White cement is available for use where white or light-colored mortar or concrete is desired. It has all the properties of regular cement and can be used in the same manner.

Early Strength Cement. This type of cement is nearly the same as ordinary portland cement except that it causes mortar or concrete to harden and develop strength much more rapidly. It can be used to advantage where construction speed is of great importance.

Air-entraining Cements. These types of cement contain small quantities of one or more chemicals that increase the amount of air held by the mortar or concrete made from the cement. The additional air tends to improve the workability of mortar or concrete and increases resistance to frost action. Air-entraining cements, while not quite so strong as ordinary or untreated portland cements, can be used to advantage in most masonry work. Air entrainment may also be achieved with air-entraining admixtures, or chemicals, added to the concrete mix.

Low-heat Cement. In cases where large volumes of concrete are placed, such as in dams or exceptionally thick walls, a great amount of heat is generated during the hardening process. This tends to cause unsightly and dangerous cracks to develop. To avoid cracking, low-heat cement was developed. It is made by decreasing the proportion of lime and using a higher percentage of silica and iron. With less lime in the cement, less heat is generated, and cracks are much less likely to develop.

Portland-Pozzolana Cement. Such cements are manufactured by grinding together portland-cement clinkers and a pozzolanic material. Rock of volcanic origin, volcanic ash, shales, and clays are among the pozzolanic materials used. Such cements improve the workability of mortar and concrete, reduce bleeding (loss of water from mortar or concrete), prevent segregation of materials, reduce heat of hydration, and increase resistance to the attack of sea water or high-sulfate soils.

Lime-Pozzolana Cements. Cements of this type are generally manufactured by grinding together hydrated lime and pozzolanic materials. Although such cements give mortar and concrete a greater resistance to sea water than standard portland cement, they set more slowly and obtain their maximum strength only after a much longer time.

Masonry Cements. Masonry cements are mixtures prepared by manufacturers to use with sand in making mortar. Such cements are generally made of portland cement and limestone but also contain other special materials which make mortar smooth and easy to work.

The American Society for Testing and Materials (ASTM) specification for masonry cement covers a type for general use. For special purposes, other masonry cements may be ordered from producers or made on the construction site.

In this book, the use of Type I standard portland cement and ASTM masonry cement is assumed.

PROPERTIES OF CEMENT

The following general information concerning the properties of cement contains interesting and instructive facts which will help readers to appreciate better how cement functions and how to use it properly and to the best advantage:

Cement Paste. Cement mixed with a little water forms a *paste*. The mixture is known as a paste because it is actually the means of binding the ingredients of mortar and concrete together, making a rocklike mass. During the manufacture of cement, all the water is driven or baked out of the materials used in making it. As long as it is kept dry, it remains in powdery or loose form. When water is added, as when making mortar or concrete, the cement forms a paste, which surrounds the other ingredients of the mortar or concrete and then hardens and grips material in contact with it.

Setting of Cement. After water has been added to cement, in either a mortar or concrete mix, the paste forms immediately, and for a short time it is plastic, that is, can be molded or shaped at will. But, as the reactions with the water proceed, the mortar or concrete begins to stiffen, or *set.* In other words, the paste in the mortar or concrete starts to assume a rocklike state. At this early stage, it is still possible, though not desirable, to disturb the mix and to reshape it without injury. However, as setting continues, the mix loses its plasticity, and if it is disturbed or reshaped, its strength will be seriously impaired. This early stage or period of hardening of cement paste is known as the *setting period.* However, there is no well-defined difference between the early and late stages of the hardening process. Once the mortar or concrete has definitely hardened, the action continues, building up a firm internal structure that gains in hardness and strength as it proceeds.

Cement paste, in either a mortar or a concrete mix, will set under water as well and as quickly as on dry land. For example, if a mortar or concrete mix is placed under water in such a manner that the cement paste cannot be separated from the mortar or concrete ingredients, either mix will set and become hard in the same manner as though on dry land. In fact, setting and hardening proceeds more satisfactorily when the mixes are kept wet or moist. This is why concrete should be kept moist during the *curing* period. The curing of concrete is explained in Chapter 4, Concrete.

Initial and Final Set. From the foregoing explanations, it is evident that, when cement is mixed with water and the ingredients of mortar or concrete, there is a time soon after the mixing has been completed when the setting period starts. As this period becomes well-advanced, it is known as *initial set.* Prior to initial set, the mixes can be disturbed and reshaped. As more time elapses, a period begins when the mixes reach a stage of *final set.* After this they cannot be disturbed or reshaped without seriously affecting their strength and dependability. Final set can.be recognized by the appearance of the mixes. If they have ceased to be plastic, they are in the final-set period and should not be disturbed in any way.

In order to avoid any possible need for disturbing mixes after the time of initial set, careful plans should be made so that they can be placed, as required, soon after the mixing has been completed. With this fact in mind, water should not be added to mixes, especially concrete, until a short time before the mixes can be placed. If concrete is purchased already mixed (see Chapter 4) and delivered by truck, the water should not be added until the truck reaches the site where the mix is to be placed. Also, the amounts of concrete placed at one time should be confined to what can be handled and finished, as explained in later chapters.

Shrinkage. As mortar or concrete hardens beyond the period of initial set, some shrinkage in its volume takes place. This is especially true in regard to mortar. For example, if a container is filled with a cement mortar mix, there is apt to be a small crack between the mortar and the sides of the container after final set has taken place. Such shrinkage is natural and cannot be avoided.

Final Set. The final or maximum strength of mixes in which regular portland cement is used as the cementitious material does not develop for a long time after the mixes have been placed. The mixes become strong enough for all practical purposes in a relatively short time. But, even after weeks and months of time, both mortar and concrete continue to become harder and stronger. If the mortar or concrete is placed under moist conditions, maximum strength will continue to develop for a very long time. Any mix gains strength at a slower rate during periods when the temperature is below 50°F.

Storage of Cement. Prior to the time cement is used, it should be stored where it can be protected against water or exceptionally moist atmosphere. It will readily absorb any moisture present and can easily be damaged or completely ruined.

Damaged Cement. If cement should, through careless storage or otherwise, absorb some moisture, it should be carefully inspected before being used. When moisture is absorbed, the cement will form into lumps. If such lumps cannot easily be pulverized by striking them lightly with a shovel or piece of wood, the cement is unfit for use. Too much emphasis cannot be placed on the danger of using damaged cement. Mortar or concrete made from such cement will not develop strength, cannot be finished properly, and constitutes a great risk so far as safety is concerned.

Volume of a Mix. Suppose, for example, that an ordinary bucket is filled with small pieces of crushed stone. Further suppose that water is poured into that bucketful of crushed stone. It is a fact that a considerable amount of water can be poured in without displacing any of the stone. There are spaces or *voids* between the pieces of stone which the water can enter. If a bucket is filled with sand, some water can also be poured in without displacing any of the sand. Even though the individual grains of sand are very small, there are voids between them which water can enter.

Earlier in this chapter, it was pointed out that cement will go through a sieve which can hold water. In other words, cement can flow into a smaller space than water can. As a further example, suppose that one bucket—level across the top—is filled with sand, and another bucket—level across the top—is filled with cement. If the two materials are thoroughly mixed, the resulting mixture will not completely fill the same two buckets. The cement enters the voids between the grains of sand and thus reduces the volume of the mix to less than two bucketfuls. This volume aspect, relative to mixes, should be kept in mind. It will be explained in more detail in Chapters 3 and 4.

CHAPTER THREE

Mortar

The earliest records of structural history show that two types of construction were employed to build walls. In one of these, walls were built by ramming together successive layers of earth. This method produced fairly strong walls which endured in climates where rainfall and dampness did not occur. In the other type, heavy blocks of stone were accurately cut and fitted, one above the other. The resulting great weight constituted the means of keeping the walls together and making them enduring. Small stone wedges were driven between blocks as a means of making tighter joints.

Somewhat later, as construction methods improved, the Egyptians invented a form of mortar which consisted of sand and a cementitious material. It is generally believed that the cementitious material was made by burning gypsum.

Still later, the Greeks and Romans improved upon Egyptian mortar by using lime and sand mixes in their mortar. Such mortar became remarkably hard and has endured through the long ages of history. In fact, examples of both Greek and Roman unit masonry still exist.

As more time went on, the Romans learned that certain volcanic deposits, if finely ground, could be mixed with lime and sand to form a mortar which was not only strong and enduring but could also resist the action of both fresh and salt water. Such mortar is still used in various parts of the world.

With improved processing of lime and the invention of portland cement, mortar materials of modern times were introduced.

14

Present-day mortar, so far as its use with concrete block and other items of unit masonry is concerned, can be defined as a mixture of cementitious materials and sand with enough water to make the mixture plastic and easy to spread and work with. As indicated by the definition, the principal components of a mortar are the cementitious materials, the sand, and a proper amount of water. Each of these ingredients serves an exact and essential function, and each should be of the best quality available. Inferior ingredients are a poor investment and never fail to cause troubles. Masonry work should be built for permanence, and this can only be accomplished by the use of first-class ingredients. The American Society for Testing and Materials publishes recommended specifications for mortar ingredients; these can be secured on request. The United States Government Printing Office, through the Superintendent of Documents, also distributes specifications recommending the best types of mortar ingredients.

The purpose of this chapter is to explain the many important aspects of mortar so that readers will understand and appreciate its functions, its desirable qualities, and the many other interesting and informative facts pertaining to it.

FUNCTIONS OF MORTAR

So far as concrete blocks are concerned, the functions of mortar can be divided into four principal groups:

1. To bind the block together so that walls and other structural details constitute a continuous unit structure which is stable and strong

2. To provide an effective barrier to the infiltration, absorption, or passing of moisture through walls and foundations

3. To prevent air and dust from passing through walls

4. To create a neat and uniform appearance

DESIRABLE QUALITIES OF MORTAR

The qualities which mortar must have in order to produce good concrete-block masonry work are set forth in the following: In most cases, first-class ingredients, when proportioned and mixed exactly as explained in this chapter, will yield mortar with satisfactory qualities. Such mortar, when placed according to the principles of good workmanship, can be depended upon.

Consistency. The consistency of a mortar refers to the amount of water in it, or its wetness. Different mortars generally require varying amounts of water to produce the same degree of wetness. The desirable consistency depends upon the way or manner in which the mortar is to be used. In most instances, masons determine the consistency which best suits their needs and add water to the mortar mixes accordingly.

Enough water must be used to yield a mortar which can be handled with ease and which will readily cover all the mortared surfaces of concrete block and properly fill all the joints.

Water Retentivity. The water retentivity of mortar is a measure of its ability to hold water. This is a most important quality because it affects almost all the other desirable qualities of mortar.

If mortar is placed on a block which is porous or has a tendency to absorb water, it may become stiff in a short time. The more porous and absorptive the block, the more readily the mortar will stiffen. A mortar which has a high water retentivity will offer greater resistance to the suction of the block and for that reason will stiffen much more slowly.

The water retentivity of mortar must be great enough so that it will remain soft long enough for a mason to place (bed) it properly before it sets. A mortar having low retentivity will stiffen before a block can be adjusted to its final position, with the result that the joint is likely to be incomplete and offer little, if any, resistance to the passage of moisture, dirt, and air through it.

If mortar stiffens too quickly when it comes in contact with an absorbent block, the next block placed on or against the mortar cannot be bedded properly in the stiffened mortar. As a result, there may be a great many cracks through which moisture, etc., can pass. In cold climates, the freezing of water in joints can destroy them almost completely.

Plasticity. This quality of mortar is closely related to consistency and water retentivity. A plastic mortar is easy to mix to a uniform composition, is readily and easily handled with a trowel, is easy to spread on a concrete block to uniform thickness, and is an aid in shoving the block into proper position. Such mortar also makes possible good bond between itself and the block.

Adhesiveness. Mortar must have enough adhesiveness to stick together and to all surfaces with which it comes in contact, to resist being squeezed out of joints, and to prevent uneven settlement from the weight of additional blocks placed above lower blocks.

Bond. The most important requirement in concrete-block construction is that the hardened mortar bind the blocks together and not just hold them apart. To make such binding (bond) possible, all contact surfaces of the blocks must have mortar applied to them. An incomplete bond offers less resistance to water infiltration. A complete bond between the mortar and the blocks helps to assure sufficient wall or foundation strength to resist stresses caused by severe winds, earthquakes, or other forces.

Volume. Concrete-block construction is apt to expand or contract a little because of temperature changes, alternate wetting and drying, or the results of water loss. If first-class ingredients are used, such volume changes are not severe enough (in some cases) to cause undesirable effects in walls or foundations. However, proper precautions should be taken to avoid them. Subsequent chapters present other ways of preventing undesirable effects.

If too little sand is used in mortar, some volume change is almost sure to take place where structural details are exposed to alternate wetting and drying. To avoid this, a somewhat larger proportion of sand should be used.

Where alternate wetting and drying do not take place very often, the proportion of sand can be reduced to form a stronger mortar; this can be done especially in relatively dry climates where earthquakes are likely.

Durability. The durability of a mortar is not always a function of its strength. For example, some of the air-entrained cements yield mortar which is not quite so strong as that made with ordinary cement as the cementitious material. Yet, the lower-strength mortar will offer greater resistance to alternate freezing and thawing. For the most part, durability depends upon a good mortar plus good workmanship.

Efflorescence. The white scum or stain which sometimes appears on walls is usually caused by certain soluble salts in the mortar, block, or water. When such salts are present, the stain is exaggerated by joints which allow the passage of water. Efflorescence can be largely avoided by using first-class mortar ingredients and good workmanship. The prepared masonry cements do not contain such soluble salts.

Color. Uniform and harmonious color is most desirable in the exposed surfaces of concrete-block walls. This can be assured if the mortar is of uniform color. The mortar color can be controlled by using the same brand of cement for all mixes and by careful proportioning to make sure that the same amounts of ingredients are used each time mixing is done. The use of prepared masonry cement also helps to keep mortar of uniform color.

KINDS OF MORTAR

For all commonly encountered concrete-block masonry work, mortar may be classified into four general types, any one of which can be used to advantage:

Straight-lime Mortar. This kind of mortar is made using either hydrated lime or lime putty, along with sand and water. Hydrated lime is a white powder which can be purchased to mix with sand and water. Lime putty has to be prepared on the job from quicklime. The process is known as *slaking.* When preparing such putty for job use, provide a clean and watertight wooden box of convenient size. Place 25 gallons of water in the box for each 100 pounds of high-calcium quicklime or 15 gallons of water for each 100 pounds of dolomitic lime. Always add the lime to the water, not the water to the lime. Have sufficient water to cover all the lime and additional water handy for quick use. Just as soon as steam forms, hoe thoroughly and add enough water to stop violent steaming. After all action has stopped, run off the resulting putty through a 20-mesh screen and allow to cool. The putty is then ready for use.

Straight-cement Mortar. This kind of mortar is made using ordinary portland cement along with sand and water.

Cement-lime Mortar. This kind of mortar is made using ordinary portland cement, hydrated lime or lime putty, sand, and water.

Masonry-cement Mortar. This kind of mortar is made using prepared masonry cement with sand and water. Mortar of this kind is becoming more and more popular because it is easy to use and because manufacturers have improved the product to the point where it can be depended upon to yield economical and satisfactory mortar.

MORTAR AGGREGATE

If mortar was made using only a cementitious material, such as portland cement, and water, the resulting mixture would shrink to such a large extent that cracks would develop in the joints between the concrete blocks that the mortar was supposed to bind together. To avoid such cracking and to make mortar much more economical in cost, sand is always mixed with the cementitious material. The sand is known as *aggregate.*

Sand. Sand must be clean to make good mortar. In other words, the sand should not contain earth or other materials. For this reason, it is always the best policy to purchase sand from a material dealer who can be depended upon to supply *pure* sand. Any percentage of earth in sand greatly reduces the strength of the mortar made from it. Thus, sand from banks should never be used.

The best mortar sand is composed of both large and small grains. The cementitious material (paste) must completely coat each grain. If the sand is composed of all fine grains, it will require more paste and the mortar will be more costly. When the sand contains both fine and coarse grains, the smaller grains fill the voids between the larger grains, thus yielding a mortar which is more workable and plastic. Also, when all sand grains are coated and lubricated with the paste of cementitious material, the smaller grains act more or less as ball bearings, thus permitting the grains of sand to roll over each other. This helps to produce uniform bedding for concrete block.

Clays. Clay is another form of aggregate which is sometimes finely ground and added to mortar as a means of making it more plastic. There is little definite knowledge available as to the performance of mortars to which such clay has been added. Therefore, in the absence of such knowledge, any form of clay should be investigated to see what its performance has been when used in mortar.

MORTAR ADMIXTURES

Use of admixtures is controversial. The Structural Clay Products Institute, in its August, 1961, Technical Notes, "Mortars for Clay Masonry," states: "Unfortunately, in recent years, it has become somewhat commonplace to use additives in mortars without regard to their effects on the hardened end product. Although some additives are harmless, some are definitely harmful. . . ."

The generally acknowledged specifications for good mortar do not accept or recognize admixtures of any kind. Thus, it would appear that admixtures should be used only when certain special properties are required in mortar.

MORTAR COLORS

Colored mortars may be prepared by the use of colored aggregate or by mixing color pigments with the cement before water is added to the mix.

Colored Aggregate. The most permanent and dependable coloring can be secured by the use of aggregate which has the natural color desired. Such mortar does not lose any of its strength and is most desirable. The coloring can be determined by mixing small trial batches and allowing them to harden. Once the proper mixture has been determined, larger batches can be mixed using the same proportions.

TABLE 1. Recommended coloring materials for mortar

Color desired	Commercial names of colors for use in cement	Pounds of color required for each bag of cement to secure	
		Light shade	Mediun shade
Grays, blue black, and black	Germantown lampblack*	½	1
	Carbon black	½	1
	Black oxide of manganese	1	2
	Mineral black	1	2
Blue	Ultramarine blue	5	9
Brownish red to dull brick red	Red oxide of iron	5	9
Bright red to vermilion	Mineral turkey red	5	9
Red sandstone to purplish red	Indian red	5	9
Brown to reddish brown	Metallic brown (oxide)	5	9
Buff, colonial tint, and yellow	Yellow ocher	5	9
	Yellow oxide	2	4
Green	Chromium oxide	5	9
	Greenish-blue ultramarine	6	9

* Only first-quality lampblack should be used. Carbon black is of light weight and requires thorough mixing. Black oxide or mineral black is probably most advantageous for general use. For black, use 11 pounds of oxide for each bag of cement.

Color Pigments. Color pigment must be resistant to the chemical action of whatever cementitious material is used. For this reason the so-called *earth pigments*, such as red, black, yellow, and brown oxides, manganese black, chromium oxide green, ultramarine blue, and carbon black, are the pigments most commonly used. Organic colors and dyes do not generally yield permanent colors in mortars. Where cost is not of great consideration, best results are usually obtained when white cement, white lime, and white sand are used. This procedure makes purer colors.

Table 1 shows colors, commercial names of colors, and the amount of each to use when making colored mortar.

When color pigments are used, they should be mixed with the cementitious materials prior to the time the sand is added to the mix. Small trial batches should be made first because the dried and hardened mortar will be lighter in color than when it is in the wet stage.

MORTAR MIXES

For concrete-block masonry, the commonly used proportion of cementitious material to sand is one part of cementitious material to three parts of sand. The proportioning or measuring is done by volume. In other words, the cubic foot is the unit of measure employed. This proportion may vary slightly, depending upon the sand. If the sand is composed of all very small grains, the amount of cementitious material must be increased. In general, the amount of sand should be near the maximum which will yield a plastic and readily workable mortar.

Measuring. As previously indicated, the required volumes of cementitious materials and sand are generally reckoned in terms of cubic feet. For this purpose, it is assumed that a bag of cement contains 1 cubic foot and that it weighs 94 pounds. Owing to the fact that various portland cements differ slightly in densities, some brands weigh a little less than 94 pounds per bag. The standard bag of hydrated lime contains 50 pounds, but the volume is generally more than 1 cubic foot. Therefore, if the cement and lime are proportioned on a volume or cubic-foot basis, the actual quantity proportions may vary considerably.

There are two possible methods for avoiding such variations. One is actually to weigh the cementitious materials. A second and simpler procedure is to proportion by bag ratios. On this basis, a $1:1:6$ cement-lime mortar (1 bag of cement, 1 bag of lime, and 6 cubic feet of sand) is prepared by mixing the materials in the ratio of 1 bag of cement, 1 bag of hydrated lime, and 6 cubic feet of sand. Mortars thus proportioned always contain cement and lime in the same weight ratios. In fact, this is the recommended procedure.

When mortars are to be made using a prepared masonry cement, the volume of sand in the mortar should be at least twice, but not more than three times, the volume of masonry cement.

Small Batches. When small batches which require less than 1 bag of cement are needed, the proportioning can be approximately accomplished by using a container, such as a bucket or an even smaller can or box. Suppose, for example, that a small batch of 1:1:6 mortar is required. Further suppose that a bucket is to be used for measuring. Measure out 1 bucket of cement, 1 bucket of lime, and 6 buckets of sand. In many cases, however, it will be more convenient to buy and use packaged masonry mortar when a small batch is needed.

Straight-cement and Straight-lime Mortars. Straight-cement mortars, that is, mortars without lime, and straight-lime mortars, that is, mortars without cement, are not often used. A straight-cement mortar would yield a mortar which is not easily workable and which would not have all the desired qualities previously set forth. A straight-lime mortar would be plastic and easily workable but would set too slowly for convenience. However, either type can be used.

STRAIGHT-CEMENT MORTAR

Use one volume of cement to three volumes of sand and enough water to make the mix workable.

STRAIGHT-LIME MORTAR

Use one volume of hydrated lime or lime putty to three volumes of sand, and enough water to make the mix workable.

RECOMMENDED MIXES

The Portland Cement Association recommends the mortar mixes shown in Table 2. These recommendations are based upon long experi-

TABLE 2. Recommended mortar mixes proportioned by volume

Type of service	Cement	Hydrated lime or lime putty, cubic feet	Damp loose sands, cubic feet
A. For ordinary service	1 masonry cement	. .	3
	or		
	1 portland cement	1	6
B. Subject to extremely heavy loads, violent winds, earthquakes, or severe frost action	1 masonry cement plus 1 portland cement	. .	6
Isolated piers	1 portland cement	¼	3

ence and experimental work. They can be followed with full confidence. For any of the mixes, use enough water to make them easily workable but not watery. The water should be fit to drink.

STRENGTH OF MORTAR

The strength a mortar must have depends upon the type of service a concrete-block wall or foundation is to give. For example, if a wall will be subject to extremely heavy loads, severe weather, or earthquakes, the mortar mix shown at B in Table 2 should be used. On the other hand, the mix shown at A is good enough for all ordinary service.

MIXING MORTAR

Mixing is an important link in the preparation of good mortar. Good sand and cement must be mixed in such a manner as to have the cement distributed evenly through every portion of the mixture, and the maximum quantity of water consistent with satisfactory plasticity must be just as evenly distributed throughout the mass.

Hand Mixing. The most common method of hand mixing mortars is by the use of the mason's hoe in a mortar box. This method continues in use only for small jobs where very small amounts of mortar are desired.

Hand mixing requires that the sand be spread in the box and the cement spread on top of the sand. The two should be mixed together by hoeing from end to end of the box until an even color is obtained. Then the mixture is spread, about two-thirds the required water is added as a pool in the center, and the hoeing process continued. As the mixing approaches completion, more water is added and the end-to-end hoeing continued, as required, to produce the necessary workable consistency.

Machine Mixing. Machine mixing can be accomplished by any number of excellent mechanical mixers available. They are similar to, but lighter than, regular concrete mixers. Their use enables large jobs to be performed more quickly, and they produce uniform, well-mixed mortars. In machine mixing, approximately half the required water should be added, followed by about one-half the required sand as the mixing action continues. Then the cement and remainder of the sand should be added. As mixing continues and the mass stiffens, the balance of the water should be added. The mixing action should be continued for at least 3 minutes after all sand and cement have been added. A total mixing time of 5 minutes is recommended.

RETEMPERING MORTAR

Retempering may be accomplished within an hour or two after mixing without damage to the ultimate properties of the mortar. Small amounts of water can be added and mixing carried on as before. The amount of extra water added should be carefully considered so that the mortar does not become watery. Add no more water than necessary to give the mortar the qualities previously explained. It is only *wet* (not watery) mortar which can develop good bond with masonry units which are absorptive. Therefore, retempering should be done as often as seems necessary to keep mortar in proper condition. When possible, the size of batches should be confined to the amounts which can be used without the necessity for retempering.

The time mortar will retain its full plasticity prior to the need for retempering depends upon the natural setting characteristics of the cement used, the temperature of the mixture, and the temperature and relative humidity of the surrounding air. Wind and sunshine also affect the situation. Mortar will retain its plasticity longest on still, damp, cool, cloudy days. Plasticity will disappear most rapidly on hot, dry, and windy days.

Mortar which has been left for several hours, such as overnight, should not be retempered or used. Mortar retempered 3 hours after mixing may lose about 25 per cent of its compressive strength.

PLACING OF MORTAR

The skill and workmanship with which mortar is placed means a great deal so far as successful, well-bonded, water-repellent, and crack-free concrete-block masonry is concerned. When proper consideration is given to the fact that each block is placed by hand, the importance of the manual operation is clearly evident.

MIXING DURING COLD WEATHER

Mortar can be mixed during cold weather just as well as at any other time if proper precautions are taken. Both the water and the sand can be heated to about 80°F. The heated water keeps the mortar warm to prevent unduly delayed setting. Heating the sand removes all ice and thaws it out. Excessive heat in the mortar can cause flash (very quick) setting because chemical reactions are faster at higher temperatures than they are at lower temperatures.

The use of an admixture to lower the freezing point of mortar should be avoided. The quantity of such materials necessary to lower the freezing point of mortar to any appreciable degree would be so large that the mortar strength and other desirable qualities would be seriously impaired. Calcium chloride is sometimes added to a mortar mix to accelerate hardening. This shortens the time required for a mortar to attain sufficient strength to resist freezing action. Generally, not more than 2 per cent of calcium chloride by weight of cementitious material should be used for such a purpose.

ESTIMATING

One cubic foot (a container which is 12 inches deep, 12 inches wide, and 12 inches long) of ordinary sand with the usual proportions of cementitious material and water will yield approximately one cubic foot of mortar. This bears out the previously explained fact that the cementitious material and water fill the voids between the grains of sand without appreciably increasing its bulk. When estimating the amounts of material required for any given project, this fact must be kept in mind.

Quantities of Materials for Small Batches. In Table 2, the recommended mortar mixes are shown in terms of proportion by volume. Table 3 indicates the same mortar mixes in terms of quantities of materials per cubic foot of mortar.

In order to mix small quantities of mortar, such as 1 cubic foot, the amount of cementitious materials and sand must be measured in terms of parts of bags and cubic feet. For example, 1 cubic foot of masonry-cement mortar (1:3 mix) requires 0.33 bag of cement and 0.99 cubic foot of sand. In like manner, 1 cubic foot of cement-lime mortar (1:1:6 mix) requires 0.16 bag of cement, 0.16 bag of lime, and 0.97 cubic foot of sand. Such fractional quantities are difficult to measure exactly. Thus an approximate measuring method must be used.

Thirty-three-hundredths (0.33) is the same as $\frac{33}{100}$ or is $\frac{1}{3}$. Thus, $\frac{1}{3}$ of a bag can be measured without too much trouble. A more accurate method is to weigh the cement. The weight of masonry cement is indicated on the bags. Thus, 0.33 bag equals $\frac{1}{3}$ the weight of a bag.

Sixteen-hundredths (0.16) is the same as $\frac{16}{100}$ or approximately $\frac{1}{6}$.

Thus, ⅙ of a bag can be measured. A more accurate method is to weigh the cement. One bag of cement weighs 94 pounds. Thus, 0.16 equals ⅙ of 94, or about 16 pounds.

For measuring sand, a No. 2 shovel can be used to advantage. For example, 5½ shovelfuls are equal to 1 cubic foot. The quantities 0.99 and 0.97 are to all practical purposes 1 cubic foot. Thus, 5½ shovelfuls of sand should be used. For the quantity 0.86, use about 4 shovelfuls.

For such small quantities, however, use of packaged masonry mortar would be more convenient. A bag of packaged mortar has a volume of about 1 cubic foot.

Quantities of Materials for Large Batches. For larger batches the quantities shown in Table 3 can be multiplied by the number of cubic

TABLE 3. Quantities of materials per cubic foot of mortar

Mortar mixes (volume)			Quantities			
Cement, bags	Hydrated lime or lime putty, cubic feet	Sand in damp, loose condition, cubic feet	Masonry cement, bags	Portland cement, bags	Hydrated lime or lime putty, cubic feet	Sand, cubic feet
1 masonry cement	..	3	0.33	....	....	0.99
1 portland cement	1	6	....	0.16	0.16	0.97
1 masonry cement plus 1 portland cement	..	6	0.16	0.16	....	0.97
1 portland cement	¼	3	....	0.29	0.07	0.86

1 bag masonry cement or portland cement = 1 cubic foot.

feet of mortar required. For example, if a cubic yard of 1:3 masonry-cement mortar is required, the 0.33 quantity can be increased 27 times because there are 27 cubic feet in a cubic yard. Thus, 0.33 × 27 = 8.91, or approximately 9 bags of masonry cement. In like manner, the 0.99 quantity increased 27 times equals approximately 1 cubic yard.

Quantities of Concrete Block and Mortar. Table 4 shows the number of concrete blocks of various sizes, necessary for 100 square feet of wall

TABLE 4. Block and mortar for 100 square feet of block wall

Materials	Concrete block—height and length, inches		
	7⅝ by 15⅝	5 by 11¾	3⅝ by 15⅝

4-INCH WALL

Number of blocks	112.5	220	225
Mortar,* cubic feet	2.3	3.6	3.9
Mortar,* cubic yards	0.085	0.133	0.144

6-INCH WALL

Number of blocks	112.5	220	225
Mortar,* cubic feet	2.3	3.6	3.9
Mortar,* cubic yards	0.085	0.133	0.144

8-INCH WALL

Number of blocks	112.5	220	225
Mortar,* cubic feet	2.3	3.6	3.9
Mortar,* cubic yards	0.085	0.133	0.144

12-INCH WALL

Number of blocks	112.5		
Mortar,* cubic feet	2.3		
Mortar,* cubic yards	0.085		

* With face-shell mortar bedding. Ten per cent wastage included.

or foundation area. The amounts of mortar are also shown for each thickness. NOTE: Face-shell bedding is explained in Chapter 8.

Variations. There is likely to be some variation in any table of mortar quantities. Cementitious materials may vary as well as the sand. The degree of dampness and compaction of sand is never the same in various localities. Therefore, masons should experiment with the materials they have to work with, keeping in mind the desired qualities of mortar as outlined earlier in this chapter.

CODES

Most cities and towns have laws and regulations which are known as *building codes*. Such codes often specify the kinds and mixes of mortar which must be used for all structural work. Readers should investigate their local codes and cooperate with the local building inspectors.

CHAPTER FOUR

Concrete

Steady improvements in the quality and performance of portland cement, coupled with noteworthy progress in concrete technology, have accelerated the tremendous growth in the use of concrete in structural details of all kinds. As a building material, it is one of the most useful because it is strong, durable, sanitary, relatively economical, and fire-resistant. It is attractive in appearance, the upkeep is low, and it can be used in any climate.

The purpose of this chapter is to explain the practical, fundamental principles of good concrete and how it is made and used.

WHAT IS CONCRETE?

Concrete is a mixture in which a paste made of portland cement and water binds aggregate (sand and crushed stone or gravel) into a rocklike mass as the paste hardens, because of the chemical reaction between the cement and the water. A strong, durable concrete is obtained by correctly proportioning and properly mixing the ingredients so that the entire surface of every particle or grain of aggregate, from the smallest to the largest, is completely coated with the cement paste and so that spaces between aggregate particles are completely filled with the paste.

When first mixed, concrete is plastic and easily molded into practically any shape. At this stage, it may be troweled to a smooth surface, brushed to have a rough texture, or shaped into ornamental patterns. It is often placed in specially made forms to reproduce elaborate ornamentation.

The proportioning of the ingredients of concrete is often referred to as *designing the mixture*. A properly designed mix will achieve the characteristics desired in both the plastic and hardened stages.

The character of concrete is greatly influenced by the quality of the cement-water paste which binds the aggregate together. If too much water is used, the paste becomes diluted and will be too weak, after hardening, to hold the aggregate firmly together. The quality of the cement paste and, ultimately, the durability, strength, and other properties of the concrete depend on the amount of water used.

The relation of water to cement is usually referred to as the *water-cement ratio*. The higher the ratio, that is, the more water used per unit of cement, the less durable and strong will be the cement. The lower this ratio, so long as the concrete is workable, the better will be the quality of the concrete.

Concrete can be made to have any desirable degree of watertightness. It can be made to hold water or other liquids and resist the penetration of wind-driven rain. Yet for some special purposes, such as filter beds, concrete can be made porous and highly permeable.

Economy in a concrete mixture designed for durability, strength, and watertightness is effected by using no more cement paste than is required to coat all the aggregate surfaces completely and fill all the voids.

The type of concrete mixture used is determined by the purpose for which the concrete is intended.

For small jobs and minor repairs, concrete can be mixed by hand, but machine-mixed concrete ensures more uniform batches, as well as more thorough mixing of ingredients.

As previously mentioned, the hardening of concrete results from a chemical action between the cement and the water. The reaction takes a little time and requires the presence of moisture and favorable temperature. During the time hardening is taking place, some of the water in the concrete combines with the cementitious material and becomes a permanent part of the concrete. However, in order to make a plastic mixture, more water must be used than can combine with the concrete. After a time, the excess water will evaporate. This is another reason why no more water should be used than is absolutely necessary.

Various structural conditions require a better concrete than others.

For example, a foundation in a wet soil must be watertight and stronger than a like foundation in a dry soil. To make watertight concrete, less water is used. In succeeding pages, more specific recommendations are given.

KINDS OF CONCRETE

In keeping with the improvements in quality and performance of portland cement, several different kinds of concrete have been developed, all of which serve specific purposes to good advantage. The most important kinds are explained in the following:

Regular Concrete. Unless otherwise specified or explained, regular concrete is a mix composed of the ingredients which have already been mentioned. Such concrete weighs about 150 pounds per cubic foot and can be made from ingredients readily obtainable in practically all parts of the country.

Air-entraining Concrete. Air-entraining concrete contains billions of microscopic air cells per cubic foot. These tend to relieve internal pressure in the concrete mass by providing tiny chambers for the expansion of water when it freezes.

This kind of concrete is produced through the use of air-entraining portland cement. The amount of entrained air is usually between 3 and 6 per cent of the volume of the concrete, but may be varied from this as required by special conditions. Air-entraining portland cement is made by grinding small amounts of soaplike resinous or fatty materials with the normal cement clinker.

The use of this kind of cement results in concrete which (1) is highly resistant to severe frost action and cycles of wetting and drying or freezing and thawing; (2) has a high immunity to the surface scaling caused by excessive amounts of chemicals used to melt pavement ice; (3) has a remarkably high degree of workability and durability. For these reasons, the value of air-entrained concrete has become widely recognized for practically all types of large construction. The manufacturers of this

type of cement supply special instructions which should be followed
when it is to be used.

Precast Concrete. Concrete cast in some location other than its final
position, allowed to harden, and then transported and erected in its final
position is called *precast concrete*. Such concrete members can be made
at a central plant, and then shipped to building sites. Or, they can be
cast at the building site.

Of the variety of precast structural members used in house and building construction, perhaps the best known is the precast concrete joist.
These lightweight reinforced-concrete beams are made in several lengths
and thicknesses and are easily set into place to support either conventional or concrete floors.

Early Strength Concrete. Builders are sometimes required to produce
concrete having a comparatively high strength within a few days after it
is placed. This can be accomplished by using a larger quantity of normal
portland cement or by using high early strength portland cement.

Concrete having high early strength can be made with normal portland
cement by reducing the total quantity of water mixed with each bag of
cement. For example, concrete made with 4½ gallons of water per bag,
as explained later in this chapter, is about twice as strong, at 3 days, as
concrete made with 7 gallons of water per bag of cement. The lower
water-cement ratio, however, will require more cement to produce the
same consistency.

The rate of hardening can be increased by dissolving flake calcium
chloride in the water. The quantity used should not exceed 2 pounds per
bag of cement; the proper amount to use will depend upon the temperature. For temperatures below 80°F, use 2 pounds per bag. For temperatures between 80 and 90°F, use 1½ pounds per bag of cement.

Another method of obtaining high early strength is to heat the water
and the aggregate as explained on page 51.

In general, the high early strength cements now available are made
from the same kinds of materials and in the same manner as normal portland cement. They harden in the same way but more rapidly. For the
same quantity of water per bag of cement, concrete made with high early
strength portland cement will be as strong at 1 day as concrete made with

normal portland cement will be at 3 days. At 7 days the strength will be about equal to the 28-day strength of concrete made with normal portland cement.

Regardless of the cement used, to attain high early strength, concrete should be kept at a temperature of 70°F or above soon after it is placed. Concrete hardens more slowly below 70°F and is not likely to have satisfactory strength at an early age.

Reinforced Concrete. In this kind of concrete, various forms of steel are embedded in the bulk of concrete as a means of preventing cracking under stresses which concrete alone cannot resist. This subject is explained in more detail in a subsequent part of this chapter.

Prestressed Concrete. Instead of ordinary reinforcing, high-strength steel bars or cables sometimes are used to prevent concrete from cracking. These bars or cables, called *tendons,* are stretched and attached to the concrete, either at their ends with anchorage devices or by bond along their length. Thus, the tendons compress, or prestress, the concrete and keep it compressed even under loads, so that cracking cannot occur.

Watertight Concrete. The essential requirement for watertight concrete is the use of durable aggregates which are completely coated with a cement paste that resists the passage of water. Leakage through concrete, if any, is usually through the paste, and it can be prevented by having a sufficient quantity of watertight paste to coat all particles and grains of aggregate and fill all spaces between them. Watertight mixes are further explained in subsequent pages.

Lightweight Concrete. In many instances, lightweight concrete can be used to advantage, especially where great strength is not a requirement. Cinder aggregate produces a form of concrete which weighs only about 110 pounds per cubic foot, as against 150 pounds where crushed stone is used for aggregate. Cinder concrete can be used for floors, roofs, and as fireproofing but should not be used in columns, beams, etc., where strength is of great importance.

There are several forms of manufactured aggregate available, all of which produce lightweight concrete.

CONCRETE AGGREGATES

Aggregates are usually divided into coarse and fine. Crushed stone or gravel is the most common form of coarse aggregate. Sand is the most common form of fine aggregate. Coarse aggregate includes crushed stone and gravel ranging from ¼ to 1½ inches or larger in size, and fine aggregate includes all particles from very fine sand (exclusive of dust) up to those which will just pass through a screen having meshes ¼ inch square.

Sound Aggregate Essential. Aggregates which are sound, hard, and durable are best suited for use in concrete. This is a most important consideration and great care should be exercised to use only such aggregates. The soft aggregates which are flaky or which may wear away when exposed to weather should never be used.

Coarse Aggregate. Trap rock is one of the hardest, most durable, and best stones to use in making concrete. Granite and hard limestones also make good aggregate. Sandstones should be avoided unless it is known that they are exceptionally hard. Crushed stone should be composed of pieces which are square, triangular, or rectangular. Flat pieces should be avoided.

Size of Crushed Stone to Use. The size of stone to use depends upon the work at hand. For example, crushed stone up to 1½ inches or more in size may be used in a thick foundation wall or heavy footing. In thin slabs, sidewalks, driveways, curbs, etc., the largest crushed-stone size should not exceed one-third the thickness of the concrete being placed. In reinforced concrete, where there are steel rods, the largest size of crushed stone should not exceed ¾ inch.

Crushed stone should contain particles which range from ¼ inch up to the largest size which can be used in the type of work being done. The

FIGURE 1. This is how well-graded aggregate appears before and after being separated into three sizes. Reading from left to right, in the separated aggregate, ¾- to 1½-inch size, ⅜- to ¾-inch size, and ¼- to ⅜-inch size. Note how smaller pieces of aggregate fit between larger ones in the mixed aggregate. (Courtesy of Portland Cement Association.)

range in size helps to produce a stronger concrete. Figure 1 illustrates a good range in size for ordinary crushed stone.

Gravel. The term *gravel* refers to stone as it occurs naturally in gravel banks. In general, this natural material makes good concrete because it is hard and durable. The size range for use in concrete should be about the same as for crushed stone. The mixture of sand and gravel, as it occurs naturally, should not be used in concrete because it contains too much sand, much of which includes objectionable foreign matter.

Crushed Stone and Gravel Must Be Clean. In addition to being sound, hard, and durable, the best crushed stone is clean and free from fine dust, loam, clay, or vegetable matter. Such materials are objectionable because they prevent the concrete paste from binding particles together. Concrete made with dirty crushed stone or gravel hardens slowly and may never harden enough to serve its purpose. It is always the best policy to obtain crushed stone from plants where it can be cleaned and properly graded.

If gravel is available for use, it should be tested to determine its cleanliness and freedom from vegetable matter. It can be tested for dirt by placing some of it in a glass jar with water. Shake well and allow to stand until the water clears. If a layer of silt covers the gravel, it is not clean enough for use and should be washed. The washing can be accomplished by placing the gravel in a large container with water. The container can be tilted so that the water, for example, from a hose, is directed to the gravel and will run away, taking dirt with it.

Gravel can be tested for vegetable matter by adding a teaspoonful of household lye to a jar containing ½ pint of water. Pour some of the gravel into the jar and shake well. If the water becomes dark brown in color, vegetable matter is present. The gravel should be washed.

Fine Aggregate. Well-graded fine aggregate, in which the particles are not all fine or all coarse but vary from fine up to those particles which will just pass a screen having meshes ¼ inch square, is recommended. If the sand is well graded, as shown in Figure 2, the finer particles help to fill the spaces (voids) between the larger particles, thus resulting in the

FIGURE 2. Sample of well-graded sand before and after it has been separated into various sizes. Particles vary from fine up to ¼-inch, the widths of the illustration strips indicating amounts of each size. For good concrete, at least 10 per cent should pass a 50-mesh sieve. (Courtesy of Portland Cement Association.)

most economical use of the cement paste in filling the voids and binding the aggregate particles together to form a strong concrete.

Sand Must Be Clean. As explained in the discussion of coarse aggregate, sand must be clean and free from loam or vegetable matter. Few natural deposits of sand, such as bank sand, are clean enough or sufficiently free from vegetable matter.

Testing sand for loam or dirt can be carried out as explained for gravel. The vegetable-matter test can be carried out using a bottle, a sample of the sand, and a 3 per cent solution of caustic soda. The caustic soda can be made by dissolving 1 ounce of sodium hydroxide in a quart of water. Put some of the caustic soda with sand in a bottle. If, after standing for 24 hours, the solution is practically colorless, the sand is free from vegetable matter. A straw-colored solution indicates some vegetable matter but not enough to be objectionable. A dark-colored solution means that the sand contains an objectionable amount of vegetable matter and should not be used until it has been washed.

Moisture in Sand. After selecting the amount of water to be used per bag of cement (as explained in the following pages) to make a paste of the desired quality, it is necessary to take into consideration the amount of water carried by the sand.

Dry Sand. Such sand is seldom available. It is *air dry*—that is, as dry as it would be if it were spread out in a thin layer and dried in the sun.

Damp Sand. Damp sand is that which feels slightly damp to the touch but which leaves very little moisture on the hands. Such sand usually contains about ¼ gallon of water per cubic foot.

Wet Sand. This is the type of sand most generally available. It feels wet and leaves considerable moisture on the hands after being handled. Wet sand contains about ½ gallon of water per cubic foot.

Very Wet Sand. Such sand is dripping wet and contains about ¾ gallon of water per cubic foot.

Fine sand generally carries more water than coarse sand, although from appearances both might seem to be equally wet.

Measuring Moisture in Sand. To become familiar with the appearance and feel of damp, wet, or very wet sand, make the following experiment: Spread a cement bagful of the sand at hand in a thin layer on a clean dry floor inside a building and let it dry. Stir it until all surface moisture disappears and the sand flows freely.

Measure out 3 gallons of this dry sand, placing 1 gallon in each of three pans. Add 5 ounces of water to one pan, 12 ounces to the second, and 20 ounces to the third, mixing each thoroughly.

The pile containing 5 ounces of water is typical of *damp* sand; that containing 12 ounces is typical of *wet* sand; and that containing 20 ounces is *very wet*.

Sand containing 2 per cent moisture carries about ¼ gallon of water per cubic foot; 4 per cent, ½ gallon; 6 per cent, ¾ gallon; 8 per cent, 1 gallon; and 10 per cent, 1¼ gallons per cubic foot.

Water. Water used to make concrete should be fit to drink.

PROPORTIONING CONCRETE

Recommended and typical qualities (mixes) of concrete for various classes of work are listed in Table 1. This is the guide for proportioning concrete ingredients according to the *total* amount of water to mix with each bag of cement. Table 1 is based upon the following facts:

1. A cement paste made in a proportion of not more than 5 gallons of water to 1 bag of cement will produce satisfactory concrete for work subjected to severe wear, weather, or weak acid and alkali solutions. Jobs that require this kind of concrete include colored sidewalks, tennis courts, floors, etc.

2. A 6-gallon paste produces a concrete which is watertight and satisfactory when subjected to moderate wear and weather. Watertight floors, foundations, driveways, sidewalks, and swimming pools are the types of work which require concrete of this quality.

3. A 7-gallon paste will produce concrete which is satisfactory where it will not be subjected to wear, weather, or water pressure. This quality of concrete is suitable for use in footings.

TABLE 1. Recommended proportions of water to cement and suggested mixes

Kinds of work	Add U.S. gallons of water to each bag batch if sand is			Suggested mixture for batch			Material per cubic yard of concrete*		
					Aggregates			Aggregates	
	Very wet	Wet	Damp	Cement, bags	Sand, cubic feet	Stone, cubic feet	Cement, bags	Sand, cubic feet	Stone, cubic feet

5-GALLON PASTE FOR CONCRETE SUBJECTED TO SEVERE WEAR, WEATHER, OR WEAK ACID AND ALKALI SOLUTIONS

Kinds of work	Very wet	Wet	Damp	Cement, bags	Sand	Stone	Cement, bags	Sand	Stone
Colored or plain topping for heavy-wearing surfaces as in industrial plants and all other two-course work, such as pavements, walks, tennis courts, residence floors, etc.	$4\frac{1}{4}$	Average sand $4\frac{1}{2}$	$4\frac{3}{4}$	1	1	$1\frac{3}{4}$	10	10	17
				Maximum-size aggregate $\frac{3}{8}$ inch					
One-course industrial, creamery, and dairy-plant floors and all other concrete in contact with weak acid or alkali solutions.	$3\frac{3}{4}$	4	$4\frac{1}{2}$	1	$1\frac{3}{4}$	2	8	14	16
				Maximum-size aggregate $\frac{3}{4}$ inch					

6-GALLON PASTE FOR CONCRETE TO BE WATERTIGHT OR SUBJECTED TO MODERATE WEAR AND WEATHER

Kinds of work	Very wet	Wet	Damp	Cement, bags	Sand	Stone	Cement, bags	Sand	Stone
Watertight floors such as industrial-plant, basement, dairy-barn, etc.		Average sand							
Watertight foundations.									
Concrete subjected to moderate wear or frost action, such as driveways, walks, tennis courts, etc.	$4\frac{1}{4}$	5	$5\frac{1}{2}$	1	$2\frac{1}{4}$	3	$6\frac{1}{4}$	14	19
All watertight concrete for swimming and wading pools, septic tanks, storage tanks, etc.									
All base course work, such as floors, walks, drives, etc.									
All reinforced-concrete structural beams, columns, slabs, residence floors, etc.				Maximum-size aggregate $1\frac{1}{2}$ inches					

TABLE 1. Recommended proportions of water to cement and suggested mixes (*Continued*)

Kinds of work	Add U.S. gallons of water to each bag batch if sand is			Suggested mixture for batch			Materials per cubic yard of concrete*		
	Very wet	Wet	Damp	Ce-ment, bags	Aggregates		Ce-ment, bags	Aggregates	
					Sand, cubic feet	Stone, cubic feet		Sand, cubic feet	Stone, cubic feet

7-GALLON PASTE FOR CONCRETE NOT SUBJECTED TO WEAR, WEATHER, OR WATER

Kinds of work	Very wet	Wet	Damp	Cement bags	Sand	Stone	Cement bags	Sand	Stone
Foundation walls, footings, mass concrete, etc., not subjected to weather, water pressure, or other exposure.	4¾	Average sand 5½	6¼	1	2¾	4	5	14	20

Maximum-size aggregate 1½ inches

* Quantities are estimated on wet aggregates using suggested trial mixes and medium consistencies—quantities will vary according to the grading of aggregate and the workability desired.

It may be necessary to use a richer paste than is shown in the table because the concrete may be subjected to more severe conditions than are usual for a structure of that type. For example, a swimming pool ordinarily is made with a 6-gallon paste. However, the pool may be built in a place where soil water is strongly alkaline, in which case a 5-gallon paste is required.

SOURCE: Courtesy of Portland Cement Association, Chicago, Ill.

Trial Batch. After the sand to be used is classified as damp, wet, or very wet, Table 1 may be referred to in determining the trial mix for any particular job. Suppose, for example, it is desired to determine the proper mix of materials, including water, for building a concrete swimming pool. For this class of work, the concrete must be watertight and be able to withstand moderate exposure.

Table 1 specifies a 6-gallon paste for concrete to be used in a swimming pool. However, for the trial batch, 1:2¼:3, made with *damp* sand, only 5½ gallons are added at the mixer because approximately ½ gallon is confined in the 2¼ cubic feet of sand. With *wet* sand, which is the kind usually available, only 5 gallons are added.

In making a trial batch, place 1 bag of cement, 2¼ cubic feet of sand, and 3 cubic feet of crushed stone or gravel into the mixer or in a box or on a platform if the concrete is to be mixed by hand. If mixed by hand, hoe the ingredients until well mixed. Add the correct amount of water, depending upon the moisture content of the sand. In building a swimming pool, using *wet* sand, this amount is 5 gallons. If a mixer is used, mix for at least 2 minutes. If it is mixed by hand, hoe until the mixture is plastic and workable. This mixture is a *trial* batch, and by noting how the concrete handles and places, masons will know whether or not to go ahead using the proportions of the trial mix in remaining batches.

If the concrete in the trial batch is a smooth, plastic, workable mass that will place and finish well, the correct mix for the job has been determined. The trial batch can be judged by working the concrete with a shovel or a trowel. The concrete should be stiff enough to stick together yet not dry enough to be crumbly. On the other hand, if the concrete is thin enough to run, it is not suitable for use. In short, the best mix is mushy but not watery or soupy.

FIGURE 3. An example of good and workable concrete. With light troweling, all spaces between aggregate are filled. (Courtesy of Portland Cement Association.)

Concrete that places and finishes readily is known as *workable concrete*. It is the kind shown in Figure 3. In a workable mixture, there is sufficient cement paste to bind the pieces of aggregate together so they will not separate. There is also sufficient cement paste and sand to give good, smooth surfaces free from rough spots. In other words, there is just enough cement paste to fill the spaces between particles completely and to ensure a smooth, plastic mix that finishes easily. If the mix looks like that shown in Figure 4, there is not enough cement and sand to fill the spaces between the crushed stone or gravel. A mix that looks like the one shown in Figure 5 has an excess of cement and sand.

If the trial batch is not workable, change the amounts of aggregate used but *do not* change the quantity of water. For example, the trial batch of one part cement to two and one-fourth parts of sand and three parts of stone or gravel may be too stiff or too mushy or may lack smoothness and workability. Each of these conditions can be corrected by changing the amounts of aggregates.

If the trial mix is too mushy, add small amounts of sand and stone in the proportions of two and one-half parts of sand and three parts of stone until the right workability is obtained. If it is necessary to use more sand than is shown in Table 1—for instance, an extra ½ cubic foot of *wet* sand—it is important to deduct the moisture carried by the additional sand. For the trial mix of $1:2\frac{1}{4}:3$, 5 gallons of water is required for each

FIGURE 4. An example of a mix in which there is not sufficient cement and sand. (Courtesy of Portland Cement Association.)

FIGURE 5. An example of a mix in which there is an excess of cement and sand. (Courtesy of Portland Cement Association.)

bag of cement when the sand is *wet*. With the addition of ½ cubic foot more sand, making the mix $1:2\frac{3}{4}:3$, only $4\frac{3}{4}$ gallons of water should be used in the following batches.

If the trial batch is too stiff and appears crumbly, succeeding batches are mixed with less aggregates.

All this may seem like a great deal of tedious work and concern. Yet, it is the way in which the best concrete is made. In the long run, the extra care brings about economy in both materials and labor. But, most important of all, such care makes durable concrete which can be depended upon under any circumstance.

Other Mixes. The mixes suggested in Table 1 are suitable for all commonly encountered concrete work and can be employed with full confidence. However, there are other mixes, such as those given under Estimating in this chapter, which are also reliable. The inexperienced mason may use the mixes suggested in Table 1 so long as these mixes do not conflict with the building code in his locality. For small jobs, packaged concrete mixes may be more convenient than purchasing and mixing separate concrete ingredients.

CONCRETE FOR HOME REPAIRS AND IMPROVEMENT

For most home repairs and improvements, such as floors, sidewalks, driveways, play courts, garden pools, etc., the following mix of concrete ingredients is recommended: 1 bag of portland cement, 6 gallons of water, and such quantities of sand and stone as will result in a plastic and workable mixture (usually $2\frac{1}{4}$ cubic feet of sand and 3 cubic feet of stone). When properly mixed, this mixture will produce a watertight concrete highly resistant to weather and wear. If the sand available is *wet,* use only 5 gallons of water per bag of cement. It should be pointed out that concrete cannot be mixed carelessly with any expectation of good results. Readers are therefore cautioned that extreme care must be exercised if they expect to obtain good results.

If smaller batches are desired, the quantities may be reduced. For example, use $\frac{1}{2}$ bag of cement, 3 gallons of water, slightly more than 1 cubic foot of sand, and $1\frac{1}{2}$ cubic feet of stone.

MAKING ECONOMICAL CONCRETE

It is usually desirable to get as much concrete as possible from each bag of cement. Thus, the more stone and sand mixed with the cement, the more concrete will be produced.

The stiff mix contains the largest amount of stone and sand and is ordinarily the most economical from the standpoint of ingredients. It is, therefore, an advantage to make mixes as stiff as can be placed easily. However, it should be pointed out that this process can be carried to extremes. If mixes become too stiff, there is a common tendency to add more water. As has been repeatedly pointed out, this should be avoided.

MEASURING CONCRETE

Accurate measurement of all ingredients, including water, is absolutely necessary in order to assure uniform batches and to produce the best quality concrete. Careless measuring results in poor concrete of a quality far short of what can be expected when proper care in measuring is exercised. When ready-mixed concrete is used, the producer should weigh the ingredients on accurately calibrated scales. For small jobs, where hand measuring is necessary, the following methods may be used:

Measuring Boxes. Wooden or metal boxes whose interior dimensions are exactly 12 by 12 by 12 inches will serve as good 1-cubic-foot measuring devices. Such boxes can be made with handles and no bottoms so that they can be placed on the platform where mixing is to be done, filled as required, and then lifted to leave sand or stone on the platform. Cement comes in 1-cubic-foot bags and does not have to be measured in such a box unless quantities less than 1 cubic foot are desired. One-half, one-fourth, and three-fourths of a cubic foot can be measured by filling such boxes accordingly.

Shovels. When measuring is done by shovels, the person doing the mixing should check the number of shovelfuls he takes in handling exactly 1 cubic foot of sand and stone. This can be done by counting the number of shovelfuls of each ingredient required to fill a cubic-foot measuring box.

Wheelbarrow. If measuring is done with a wheelbarrow, the barrow should be marked on the inside edges for 1 or 2 cubic feet, etc. This marking can be done by dumping a cubic-foot measuring box of material into the barrow, leveling, and making a mark at that level. This can be repeated with another cubic foot of material, etc., until the barrow is properly calibrated.

Measuring Water. An ordinary 12-quart pail, marked off in gallons, and quarter gallons, can be used for measuring water if a power mixer equipped with a water-measuring device is not used for the mixing.

MIXING CONCRETE

When power mixers are employed, the ingredients for each batch should be put into the hopper or directly into the mixer and the proper amount of water added last. Mixing should continue for at least 2 minutes. When ready-mixed concrete is used, water should be added only a few minutes, at the most, before the concrete is placed.

Hand Mixing. Good mixing can be accomplished by hand if the following suggestions are kept in mind:

1. The mixing should be done in a large, shallow box or on a wood platform or other flat, smooth base which will not absorb water or allow it to leak away. Platforms are handier because they allow more freedom in the use of shovels.

2. Use a measuring box, a shovel, or a wheelbarrow to obtain the correct quantity of sand and spread it out evenly.

3. Dump the required amount of cement directly on the sand and spread it evenly.

4. Mix the sand and cement, using a shovel or a hoe, until the mixture is of uniform color and free from streaks of brown or gray.

5. Measure the stone and spread it over the sand and cement mix evenly. Mix the stone with the sand and cement. The mixture should be turned over and over several times.

6. Make a depression in the center of the mix and slowly add the proper amount of water. Turn the mix in toward the water. Finally, keep turning the mix over and over until it has the desired workability and smoothness.

7. When the work is done, carefully wash the box or platform to remove all remnants of the concrete.

NOTE: The *placing* of concrete is explained in subsequent chapters in connection with different concrete projects. For each such project refer to this chapter for the proper mix to use.

REINFORCED CONCRETE

Figure 6 shows a purely hypothetical representation of typical concrete foundations, along with footings, and a concrete beam which is to be supported at both ends by the foundations. The illustration is not drawn to scale but will serve the purpose in discussing *compressive* and *tensile* strengths so far as concrete is concerned.

Compressive Strength. By compressive strength of concrete is meant its ability, in any shape or form, to support heavy weights (called *loads*) without breaking or crushing. Concrete has great compressive strength. It can support loads of 2,000 pounds over every square inch of its surface without breaking or crushing. *However, the concrete itself must be supported in order to have such great compressive strength.*

In houses and other buildings, the foundations support the loads from the outside wall above them, plus half the loads from floors and roofs. In Figure 6, the P_1 arrows represent such loads on the foundations. The foundations are supported by the footings which are in turn supported by the soil.

When concrete is amply supported and in turn supports loads without failing, it is said to have compressive strength. It should be understood that the loads are applied straight down and not from one side or the other.

Tensile Strength. By tensile strength of concrete is meant its ability, while having most of its bulk unsupported, to resist or support heavy loads without breaking or tearing apart. Concrete has little tensile strength. In the form of a beam, such as shown in Figure 6, and in many other forms, it can support only small loads without breaking or tearing apart, and is therefore classed as having no appreciable tensile strength.

Suppose that the P_2 arrows represent loads coming down on the beam. As the loads increase beyond the concrete's small tensile strength, the beam will tend to break or tear apart at Y, near the mid-point of its underside. In other words, without tensile strength, the beam will fail.

The P_2 loads cause a *compressive* stress at the top of the beam and a *tensile* stress at the bottom. Concrete has ample compressive strength when firmly supported. But in the case of the beam, there is no support under it, near the midsection. Therefore, it has no appreciable tensile strength.

Reinforcement. Unlike concrete, steel has great tensile strength. *Reinforcement* is the term used to describe the steel bars placed in concrete to increase its tensile strength *greatly*.

In the case of the beam shown in Figure 6, all the necessary tensile strength can be added to the concrete by placing two or more steel rods

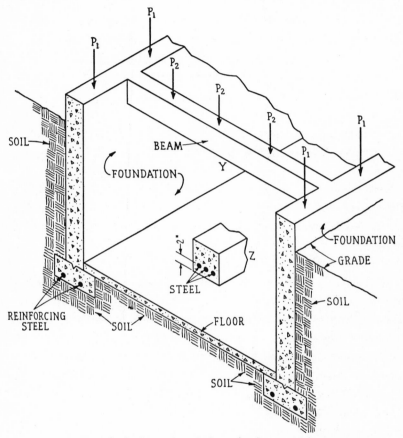

FIGURE 6. Concrete foundations, footings, and beam.

near its lower side, as shown by the insert at Z. The amount of steel and its size depend upon the loads being supported.

Sometimes, depending on the load from the foundations and the load-carrying capacity of the soil, footings also require reinforcement, as shown in Figure 6. The steel adds great tensile strength to the footing and prevents its being broken or torn apart.

For the most part, reinforcement should be designed by structural engineers, and it is therefore beyond the scope of this book. However, many building codes, especially in earthquake regions, exactly specify what reinforcement must be used. In many cases, the specifications are so detailed as to constitute all the design necessary.

In general, it is necessary that all reinforcement be protected by at least 2 inches of concrete, as shown at Z in Figure 6. The rods can be held in position, prior to the placing of concrete, by metal ties, spacers, or wire. All reinforcement should be free of dirt, rust, scale, or other coating.

Subsequent chapters explain many applications of reinforcement to houses and other small buildings.

CONCRETE IN COLD WEATHER

With proper precautions and care, concrete can be mixed and placed during cold weather as well as at any other time.

Heat hastens the hardening of concrete; cold retards it. When concrete is placed in forms, it is best that it have a temperature of not less than 60°F or more than 80°F. In cold weather, the concrete must be maintained at a temperature of 50°F or higher for at least 5 days after placing.

In early winter, when freezing temperatures usually occur only at night, it is necessary merely to protect the concrete from freezing after it is placed. During freezing weather, the water and aggregate must be heated before mixing.

Heating Water. Water can be heated in large kettles supported over a fire. Care must be taken that its temperature is never higher than 175°F when it comes in contact with the cement in the mixture; otherwise, a flash set may take place and ruin the mix. Boiling water may be added to the aggregate before the cement is added.

Heating Aggregate. Sand and stone can be heated in any number of ways. Metal barrels, old smokestacks, etc., can be used to good advantage. For small jobs, a good heater can be made by building a firebox of concrete block having a sheet-iron cover on which the aggregates can be piled.

Sand and stone should be heated separately so that they do not become improperly mixed. Cement should not be heated under any circumstances. Heat the aggregates until they feel hot to the hands but do not burn them.

Mixing and Placing Concrete. The concrete should be made as stiff as good workability allows and should be placed at once. Care should be taken to remove all snow, frost, or ice from the forms before placing the concrete.

Admixtures. The same directions as given in Chapter 3, Mortar, apply to concrete.

Protecting Concrete. After the concrete is placed, sidewalks, floors, pavements, etc., can be protected by covering them with heavy paper and a 10- to 12-inch layer of hay or straw. Walls and other vertical details can be protected by canvas sheets or by piling hay or straw around them. Interiors of buildings may be enclosed with plastic sheets, if necessary, and heated by salamanders or other heating devices.

Form Removal. During cold weather, forms should not be removed for at least a week. The longer they are left in position, the better.

Frozen Concrete. If concrete freezes before its final set has taken place, the results are apt to be most harmful. The freezing action disrupts the mass and permanently impairs strength. This is especially true in regard to thin details which do not have much bulk.

Heavier sections of frozen concrete can be reclaimed by flooding them with hot water until thawed. Care must be exercised not to disturb the forms or the mass of the concrete. Thawed concrete is not likely to acquire the full strength it would otherwise have had. Once the concrete is thawed, it should be frequently flooded with warm water and kept covered with burlap or canvas and a layer of hay or straw.

If, after forms have been removed, the concrete is crumbly, it should be junked because crumbling is a sure sign that the freezing ruined it.

CURING CONCRETE

If concrete is properly proportioned, thoroughly mixed, and carefully placed, the final step to assure watertight, durable, and strong results is to provide the right curing conditions. Curing is every bit as important as any other aspect of concrete work and should not be overlooked or slighted.

As previously explained, concrete hardens because of a chemical reaction between the cement and the water. That process continues to the best advantage when moisture is constantly present.

After concrete is placed, its desirable qualities develop very rapidly during the first week. In fact, the gain in strength during that period is as much as or more than during the following 3 months. However, if concrete becomes dry, the gain in strength stops. For this reason, it is important to keep it moist for a time after it is placed. The process of keeping it moist is known as *curing*.

Methods of Curing. After concrete is poured, especially for floors, curbs, driveways, etc., where a large surface is open to the air, the concrete surface soon starts to harden. When that initial hardening reaches the stage where no surface water is seen and when the surface cannot be marred easily, burlap or canvas can be soaked with water and spread over the concrete surface. Frequent wetting should continue over a period of from 5 to 8 days, especially during hot and dry weather. Hay or straw can also be used and should be kept wet for the same length of time. Or a sealing compound may be sprayed over the concrete surface to prevent escape of moisture.

Walls and other vertical surfaces can be protected during the curing period by leaving the forms in place or by hanging burlap or canvas over them. As previously stated, frequent wetting is necessary.

Floors and sidewalks which must have high resistance to wear should be cured with special care during the first 5 days after they have been placed. If careful curing is performed, all such defects as *dusting* and *surface checking* can be avoided.

Effect of Temperature. Concrete cures to the best advantage (if properly moist) when the temperature is between 70 and 80°F. As the temperature is lowered below 70°, the rate of hardening decreases rapidly until at 33° (just above freezing) it takes more than three times as long to develop a given strength as at 70°. Ideal temperature conditions are seldom attained but should be planned for where possible.

SPECIAL SURFACE FINISHES

A great variety of special wall finishes can be produced by using special aggregates, special forms, or special treatment after the forms have been removed.

Special Aggregate. Aggregates may be selected because of their color or for their ability to take polish. Such aggregates include white sand, marble chips, granite screenings, crushed feldspar, mica spar, slag, garnet sand, and similar colored materials. The mixtures are made and placed in the usual way. The surface finish is secured by washing off the surface film of cement, exposing the colored aggregate.

If forms are removed within 24 hours, the surface film can be washed off by spraying with water under pressure or by scrubbing with a stiff brush and water.

To produce a granolithic surface, a mix of approximately one part cement, one and one-half parts of fine aggregate, and two and one-half parts of coarse aggregate made up of pebbles, crushed granite, or other stone, as described, is used for the facing or topping. For walls, the granolithic surface is made by placing about 1 inch of facing materials against the face of the forms before the backing of ordinary concrete is placed.

Special Forms. Special finish can be imparted to concrete by using tongued-and-grooved well-fitting wood forms. The surface of the finished concrete will have whatever pattern exists in the boards.

Rubbed Finish. To produce a rubbed finish, the forms should be removed after about 24 hours. The surface of the concrete should be wetted thoroughly and then scrubbed with carborundum stones. Any lather can be removed by brushing and water. Voids in the surface can be filled with mortar. The mortar should be worked in with a carborundum stone. Several weeks after the first scrubbing, another rubbing can be performed to make the surface smoother.

Scrubbed Finish. To produce a finish of this kind, remove the forms after about 24 hours and scrub the surface of the concrete with a wire brush and water. Continue the scrubbing until the aggregate is exposed. Rinse with clear water. When scrubbed surfaces are planned, fillets of rounded molding should be placed in the forms so that no sharp corners exist.

EFFLORESCENCE

Efflorescence is apt to appear on concrete surfaces when water drains out of the material and carries with it soluble substances. Good water-tight concrete can practically prevent such scum.

If such scum does occur, it can be dissolved using a dilute solution of muriatic acid—one part acid to ten parts water. The surface should be wetted before the acid solution is applied with a brush. Allow the acid solution to remain on the concrete for about 5 minutes and then rinse with clear water.

ESTIMATING

In good concrete, where well-graded aggregate is used, much of the space occupied by a mass of concrete is taken up by the stone; the spaces (voids) between the particles of stone are filled with sand; the cement fills the spaces between the grains of sand. Because the sand and cement fill the spaces between the stone particles, a mass of concrete will actually occupy less space than the three ingredients separately. For example, a concrete mix composed of 1 cubic foot of cement, 2 cubic feet of sand, and 4 cubic feet of stone totals 7 cubic feet before mixing. After mixing, the ingredients make a mass which is not much more than 4 cubic feet. As a further example, note the 1:1:1¾ mix shown in Table 1. For 1 cubic yard, 10 bags of cement (10 cubic feet), 10 cubic feet of sand, and

TABLE 2. Materials required for 1 cubic yard of mortar and concrete

Mixture	3/4-inch gravel			1-inch stone,* dust out			2½-inch stone,† dust out			2½-inch stone, most small stone out		
	Cement, bags	Sand, cubic yards	Stone, cubic yards	Cement, bags	Sand, cubic yards	Stone, cubic yards	Cement, bags	Sand, cubic yards	Stone, cubic yards	Cement, bags	Sand, cubic yards	Stone, cubic yards
Mortar												
1:1½	14.5	0.80	....	....	....	....	15.5	0.86				
1:2	12.0	0.90	....	....	....	....	13.0	0.95				
1:2½	10.5	0.96	....	....	....	....	11.0	1.01				
1:3	9.0	1.01	....	....	....	....	9.5	1.06				
Concrete												
1:1:2	9.0	0.35	0.74	10.5	0.39	0.78	10.5	0.40	0.80	11.0	0.41	0.83
1:1½:3	7.0	0.39	0.78	7.5	0.42	0.84	7.5	0.43	0.87	8.0	0.45	0.89
1:1¾:2¾	7.0	0.43	0.75	7.5	0.47	0.80	7.5	0.46	0.84	8.0	0.48	0.85
1:2:3	6.0	0.47	0.73	7.0	0.52	0.77	7.0	0.53	0.79	7.0	0.54	0.81
1:2:3½	6.0	0.44	0.77	6.5	0.48	0.83	6.5	0.49	0.85	6.5	0.50	0.88
1:2:4	5.5	0.41	0.81	6.0	0.44	0.89	6.0	0.45	0.90	6.0	0.47	0.93
1:2½:4	5.0	0.47	0.75	5.5	0.52	0.82	5.5	0.53	0.84	5.5	0.54	0.87
1:2½:4½	4.5	0.44	0.80	5.0	0.48	0.87	5.0	0.49	0.88	5.5	0.51	0.91
1:2½:5	4.5	0.42	0.83	5.0	0.46	0.91	5.0	0.46	0.92	5.0	0.48	0.96
1:3:4	4.5	0.52	0.72	5.0	0.58	0.77	5.0	0.58	0.78	5.5	0.60	0.80
1:3:5	4.0	0.47	0.78	4.5	0.51	0.85	4.5	0.52	0.87	4.5	0.54	0.89
1:3:6	3.5	0.42	0.84	4.0	0.46	0.92	4.0	0.47	0.93	4.0	0.48	0.97

* Using very fine sand.
† Using coarse sand.
SOURCE: Courtesy of *Practical Builder*, Chicago, Ill.

TABLE 3. Volume factors of various mixes

Kind of concrete work	Mix by volume job, damp materials			Work-ability or consistency	A 1-bag batch makes this volume of concrete, cubic feet	Total water per bag, gallons	Materials for 1 cubic yard of concrete			Materials for 100 square feet, 1 inch thick		
	Cement, bags	Sand, cubic feet	Stone gravel, cubic feet				Cement, bags	Sand, cubic feet	Stone gravel, cubic feet	Cement, bags	Sand, cubic feet	Stone gravel, cubic feet
Footings, heavy foundations	1	3.75	5	Stiff	6.2	8.00	4.3	16.3	21.7	1.34	5.02	6.71
Watertight concrete for cellar walls and walls above ground	1	2.5	3.5	Medium	4.5	6.00	6.0	15.0	21.0	1.85	4.63	6.48
One-course driveways, floors, walks	1	2.5	3	Stiff	4.1	5.50	6.5	16.3	19.5	2.03	5.09	6.08
Two-course driveways, floors, walks	1	Top, 2	0	Stiff	2.14	....	12.6	25.2	....	3.89	7.78	7.01
	1	Base, 2.5	4	Stiff	4.8	6.00	5.7	14.2	22.8	1.75	4.38	6.91
Pavements	1	2.2	3.5	Stiff	4.2	5.25	6.4	14.1	22.4	1.98	4.35	
Watertight concrete for tanks, cisterns, and precast units (piles, posts, thin reinforced slabs, etc.)	1	2	3	Medium	3.8	5.00	7.1	14.2	21.3	2.18	4.35	6.54
	:	....	...	Wet	3.9	5.75	6.9	13.8	20.7	2.13	4.26	6.39
Heavy-duty floors	1	1.25	2	Stiff	2.8	....	9.8	12.3	19.6	3.03	3.79	6.06

SOURCE: Courtesy of *Practical Builder*, Chicago, Ill.

TABLE 4. Materials for 100 square feet of waterproofed walls, floors, sidewalks, or slabs

CONCRETE BASE

Slab thickness, inches	1:1¾:2¾			1:2:3			1:2:3½			1:2½:4			1:3:5		
	Cement, bags	Sand, cubic yards	Stone, cubic yards	Cement, bags	Sand, cubic yards	Stone, cubic yards	Cement, bags	Sand, cubic yards	Stone, cubic yards	Cement, bags	Sand, cubic yards	Stone, cubic yards	Cement, bags	Sand, cubic yards	Stone, cubic yards
2½	5.7	0.36	0.62	5.2	0.40	0.59	4.8	0.37	0.64	4.2	0.40	0.63	3.4	0.39	0.65
3	6.8	0.43	0.74	6.3	0.48	0.71	5.8	0.44	0.76	5.0	0.48	0.75	4.1	0.47	0.78
3½	8.0	0.51	0.86	7.3	0.56	0.83	6.8	0.52	0.90	5.8	0.56	0.88	4.8	0.55	0.92
4	9.1	0.58	0.99	8.4	0.64	0.95	7.7	0.59	1.02	6.6	0.64	1.01	5.5	0.63	1.05
4½	10.3	0.65	1.11	9.4	0.72	1.06	8.7	0.66	1.15	7.5	0.72	1.13	6.1	0.70	1.17
5	11.4	0.73	1.23	10.5	0.80	1.19	9.7	0.74	1.28	8.3	0.80	1.26	6.8	0.79	1.31
5½	12.6	0.80	1.36	11.6	0.88	1.31	10.7	0.82	1.41	9.2	0.88	1.39	7.5	0.87	1.45
6	13.7	0.87	1.48	12.6	0.96	1.42	11.6	0.89	1.54	10.0	0.96	1.52	8.2	0.94	1.57

WEARING OR FINISH COURSE

Thickness, inches	1:1½		1:2		1:1:1			1:1:1½			1:1:2		
	Cement, bags	Sand, cubic yards	Cement, bags	Sand, cubic yards	Cement, bags	Sand, cubic yards	Stone, cubic yards	Cement, bags	Sand, cubic yards	Stone, cubic yards	Cement, bags	Sand, cubic yards	Stone, cubic yards
½	2.4	0.13	2.0	0.15	2.1	0.08	0.08	1.8	0.07	0.08	1.6	0.06	0.12
¾	3.6	0.19	2.9	0.22	3.1	0.11	0.11	2.7	0.10	0.11	2.4	0.09	0.18
1	4.8	0.26	3.9	0.29	4.2	0.15	0.15	3.7	0.14	0.15	3.2	0.12	0.24
1¼	6.0	0.33	4.9	0.36	5.2	0.19	0.19	4.6	0.17	0.19	4.1	0.15	0.30
1½	7.2	0.40	5.9	0.43	6.3	0.23	0.23	5.5	0.20	0.23	4.9	0.18	0.36
1¾	8.4	0.46	6.9	0.50	7.3	0.27	0.27	6.4	0.24	0.27	5.7	0.21	0.42
2	9.6	0.53	7.9	0.58	8.3	0.31	0.31	7.3	0.27	0.31	6.5	0.25	0.50

SOURCE: Courtesy of *Practical Builder*, Chicago, Ill.

17 cubic feet of stone are required. This makes a total of 37 cubic feet. There are 27 cubic feet in a cubic yard. Thus, the ingredients, before mixing, occupy 10 more cubic feet than after mixing. These facts must be reckoned with when estimating the amount of concrete necessary for any purpose.

Table 1 shows the proportions for several trial mixes, anyone of which can be used as indicated. The table also shows the amounts of ingredients, for the four proportions, necessary to make 1 cubic yard of concrete.

Table 2 shows the proportions for several other recognized mixes, according to the size of stone used and the ingredients necessary for 1 cubic yard of concrete.

Table 3 shows the volume factors for several other often-used proportions, together with other useful estimating information.

Table 4 shows the amount of ingredients, in terms of common proportions, required for 100 square feet of walls, floors, sidewalks, and slabs of various thicknesses.

CHAPTER FIVE

Concrete Footings

In Chapter 4, Concrete, it was pointed out that the great compressive strength of any mass of concrete is available and dependable only when the concrete is properly supported. This fact is illustrated in subsequent examples pertaining to foundations. (The principles involved and methods of construction of foundations are explained in Chapter 6, Concrete Foundations.) Any foundations which are well planned and constructed are capable of supporting great weights (loads) so long as they are, in turn, properly supported. In order to assure such proper support, *footings* should be employed under the foundations.

Many other structural items also require the support which footings provide. For example, a column (a vertical structural member used to support one or more horizontal structural members) will safely support great loads if it is, in turn, properly supported.

Footings are used only when structural members indirectly depend upon the soil (earth) for support. Foundations extend down into the soil and are, by the aid of *footings,* supported by the soil. A column of the sort seen in the basement of a house is, by the aid of a footing, supported by the soil.

It is probable that few inexperienced masons fully realize the importance of footings or appreciate the care which should be exercised in their planning and construction. This fact cannot be emphasized too strongly. Unless footings are given the construction care they merit, any building they are a part of may develop cracks, off-level floors, sticking doors and windows, and many other undesirable ailments, including possible danger of collapse.

The purpose of this chapter is to explain the principles of footings, illustrate the commonly encountered kinds and their uses, set forth good planning procedure, describe how to build footings, and present many other important aspects of their use.

STRENGTH OF SOILS

This subject is discussed first because of its great importance in practically all matters pertaining to the planning of footings. As previously mentioned, soils actually support many structural details which, in turn,

TABLE 1. Load-carrying capacities of soils

Type of soil	Capacity, tons per square foot
Soft clay	1
Wet sand or firm clay	2
Fine, dry sand	3
Hard, dry clay or coarse sand	4
Gravel	6
Hardpan or shale	10
Solid rock	*No limit*

support other structural members. Thus, it is wise to give the subject of soil strength careful consideration.

Various soils, such as soft clay and dry sand, differ greatly in their ability to support load-bearing structural members, such as foundations or columns, in such a manner that the structural members will not sink or settle down into them. Engineers refer to such ability as the *load-bearing* or *load-carrying capacity* of the soils, it being understood that the term means the same as the ability of soils to resist the sinking or settlement of any structural detail.

Table 1 shows the load-carrying capacities of several commonly encountered soils. For example, soft clay can support 1 ton of weight per square foot. This means that, if a structural member has a base (or footing) equal to 1 square foot and carries or supports a load of 1 ton, it would not sink or settle in the clay. Dry sand has a greater load-carrying capacity than soft clay. For example, if a structural member has a base (or footing) equal to 1 square foot and carries or supports a load of 3 tons, it would not sink or settle in the sand.

The type of soil present at any proposed building site can be determined by careful inspection. Table 1 can be used to estimate the strength the soil has and what loads per square foot it can safely support without the possibility that structural members will sink or settle. When there is any doubt about the type of soil present, it is the wisest policy to assume the load-carrying capacity of soft clay, or better still, to make load tests.

In order to explain the purpose of footings so that their general principles can be visualized and understood, two typical examples are presented in the following:

Example 1. The *X* part of Figure 1 shows a picturelike view of part of a typical foundation. It is 12 inches thick, 8′ 0″ high, and supported by soft clay. The weight (load) which it must support comes from the walls, floors, and roof above it. Figure 1 shows only a portion of such structural members to illustrate the point.

When considering the strength of a foundation or its ability to support safely the loads previously mentioned, engineers consider only 1 *lineal foot* of it. This is because a distance of 1 foot is easier to work with in terms of the load. Then, too, if 1 lineal foot is amply strong, the whole foundation will be amply strong. The dashed lines, lettered *ABCD* at their lower ends, represent 1 lineal foot of the foundation. Just that 1 lineal foot will be considered.

An ordinary two-story frame house, along with the weight of its furniture, equipment, and occupants, might constitute a weight equal to about 2,500 pounds *per lineal foot* on the foundation under it. In other words, the lineal foot marked *ABCD* in Figure 1 might have to support about 2,500 pounds.

The foundation itself has considerable weight because concrete weighs 150 pounds per cubic foot. The total cubic feet in 1 lineal foot of the foundation shown in Figure 1 is equal to 1 foot (length) times 1 foot (width) times 8 feet (height), or 8 cubic feet. Then, $150 \times 8 = 1,200$ pounds which is the weight of 1 lineal foot of the foundation. The total weight exerted on the soil by 1 lineal foot of the foundation is, therefore, $2,500 + 1,200 = 3,700$ pounds, or $3,700/2,000 = 1.85$ tons.

Distribution of Foundation Load. The bottom surface (*ABCD*) of 1 lineal foot of the foundation is 1 foot square. From Table 1, 1 square foot of soft clay can carry only 1 ton. Thus, it cannot support 1 lineal foot of the foundation. Therefore (and this is where footings enter the picture) the load per lineal foot must be spread over a greater area. In order to determine the required area, 1.85 tons must be divided by 1 ton. This

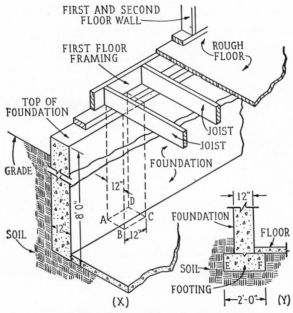

FIGURE 1. Foundation with and without footing.

calculation gives 1.85 square feet. Since 1.85 is almost 2 square feet, it can be assumed that the load per lineal foot of the foundation must be distributed over that area. Thus, the load per lineal foot of the foundation, when distributed over 2 square feet of surface, amounts to one-half of 1.85, or 0.925 tons per square foot. This is less than the maximum load-carrying capacity of soft clay of 1 ton per square foot.

In order to actually distribute the load over 2 square feet, a footing such as shown in the *Y* part of Figure 1 should be provided to support the foundation. The foundation will then be supported by the footing, and the footing will be supported by the soil. The footing is 2′ 0″ wide. Thus, every lineal foot of it constitutes an area of 2 square feet.

If the foundation shown in Figure 1 was in a soil consisting of fine, dry sand, no footing would be necessary because such soil has a load-carrying capacity of 3 tons, or 6,000 pounds per square foot. However, most city and town building codes require footings regardless of the soil condition. Use of a footing makes sure that cracks and other undesirable aspects of sinking or settlement will never occur.

Example 2. The foregoing example applies equally well to a foundation constructed of concrete block. However, because such foundations are composed of units which are held together by mortar joints, footings are also required as a means of keeping them all together as a whole. Without footings a concrete-block foundation, even if reinforced with steel, would be apt to crack at the mortar joints and thus lose much of its strength and weather-tightness. Therefore, under all circumstances, footings *must* be used with such foundations.

KINDS OF FOOTINGS

There is a great variety of footings, some of which are used for special purposes, while others are merely shaped differently to serve the general purposes explained previously. The most commonly used kinds are explained in the following:

(A) FIGURE 2. This is the most popular kind of footing and is used for foundations in houses, apartment buildings, and other medium-sized structures. It serves its purpose equally well for both concrete and concrete block. This is the kind of footing required in Plate VI.*

(B) FIGURE 2. Quite often the total load encountered in porches, garages, and other one-story frame buildings is light enough so that adequate support for foundations is provided by this kind of a footing, which is placed as an integral part of the foundations. Building codes should be checked before planning to use such a footing.

(C) FIGURE 2. This kind of footing, which is also placed as an integral part of foundations, can be used for lightweight buildings, especially of the types built for farm use. The use of this kind of footing depends upon

* House plans, Plates I to VI, appear at the end of the book.

whether the soil is such that the sides of the excavation will stand erect and serve the same purpose as forms.

(D) FIGURE 2. Columns, such as those to support girders and beams, exert great weight at their bases. In many instances they carry concen-

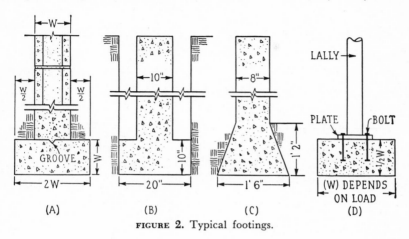

FIGURE 2. Typical footings.

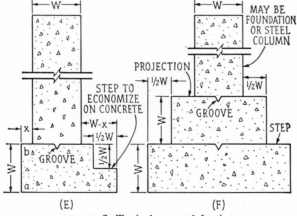

FIGURE 3. Typical stepped footings.

trated loads from large areas of floors and roofs. Generally, their bases or base plates are no more than 1 foot square. In most soils, a footing having an area of at least 4 square feet is required for such columns.

(E) FIGURE 3. This kind of footing is frequently employed when a building is being erected so close to a property line that the foundation cannot be centered on it. The side of such a footing, for example, *ab*, is placed at the lot line. In order to bring about some concrete economy, a step can be made without impairing the strength of the footing.

(F) FIGURE 3. Stepped footings can be used with either foundations or columns. When the bottom width of footings has to be over 2′ 0″ wide, the stepped construction makes for economy in the use of concrete.

(G) FIGURE 4. In many instances, precast concrete piers are used for buildings which have crawl spaces under them. Such piers are held at the proper level by means of 2 by 4 stakes while the concrete footings are placed around and under them.

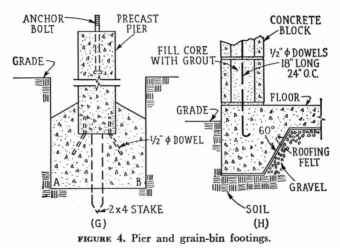

FIGURE 4. Pier and grain-bin footings.

(H) FIGURE 4. The grain-bin type of footing, while originated for such buildings, can also be used in connection with concrete floor slabs in sections of the country where frost and dampness do not have to be considered. Either frame or masonry exterior walls may be used with such footings.

(I) FIGURE 5. In dry and frost-free climates, this kind of footing can be used to advantage, especially where concrete-block walls are concerned. The footing and floor are placed integrally and thus offer resistance to unusual shocks and other stresses.

(J) FIGURE 5. This type of footing is a variation of the previously mentioned kind. The only difference is that the footing is not an integral part of the floor. Also, the first course of blocks acts as a foundation for all other courses above.

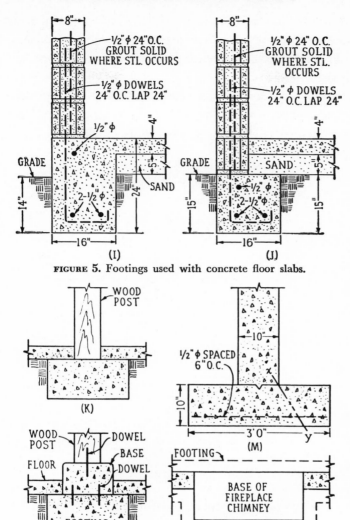

FIGURE 5. Footings used with concrete floor slabs.

FIGURE 6. Special footings.

(K) FIGURE 6. When wood posts are used as columns to support girders or beams, the required footing may be similar to the one shown at *D* in Figure 2 except that the concrete floor acts to keep the column base from slipping or sliding. While this kind of a column footing is satisfactory where no dampness is apt to occur, there is a better kind which can be made with some additional expense.

(L) FIGURE 6. In this kind of column footing, an additional concrete base is employed in order to keep the bottoms of wood columns above any possible moisture. Dowels must be used to anchor both the concrete base and the column.

(M) FIGURE 6. In cases where exceptionally wide footings (over 2' 0") are required and where steps are not desirable or possible, the footings must be reinforced as a means of preventing cracks along such lines as shown at *xy*.

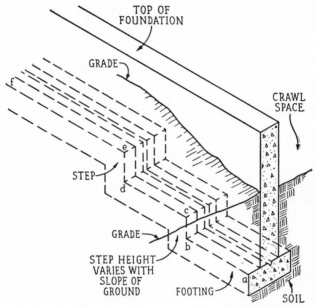

FIGURE 7. Footings for foundations in sloping soil.

(N) FIGURE 6. All fireplaces and regular chimneys must have footings of sufficient area to distribute their weights according to the soil condition. Such footings are placed as an integral part of the foundation footings when the fireplaces and chimneys are in exterior walls, or they are similar to column footings when the fireplaces and chimneys are separate from the foundations.

Footings in Sloping Soil. When houses are built on sites which slope, such as a hillside, the footings often have to be stepped, as shown in Figure 7. Such footings can be used to advantage when crawl spaces are planned or when only a partial basement is desired.

FOOTINGS MUST BE SQUARE AND LEVEL

One of the most important aspects of footings has to do with the shapes of their sides and bottoms. There are several principles which should be understood and carefully followed in building footings so as to avoid any one of several unfortunate situations.

Shapes. When excavations for footings are made, great care should be taken to remove enough soil to provide full depth and level bottoms. Then, forms should be used, except for the kinds of footings shown in the

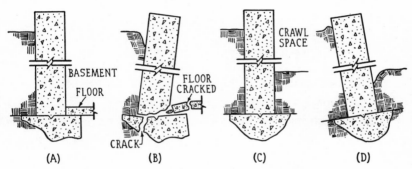

FIGURE 8. Poorly shaped footings and possible failures.

G part of Figure 4. Unless footings have full depth, level bottoms, and square sides, they cannot transmit loads to the soil safely and properly.

The *A* part of Figure 8 shows a poorly shaped footing for a basement foundation. This footing lacks full depth and its base is uneven. The *B* part of the same illustration shows what could happen to such a footing and to the foundation and floor above and to one side of it.

The *C* part of Figure 8 shows another poorly shaped footing, and the *D* part indicates what could happen as a result of such an inadequate footing. The ball-like base of such a footing might crack or cause a severe tilt in the foundation.

If any of the foregoing failures take place, adjoining parts of foundations may be seriously affected. Because of poor support under one part of a foundation, other parts could quickly lose their strength, crack, or otherwise fail.

Level. The entire length of footings, such as those required for the house indicated in Plates I through VI, must be at the same level unless stepped. Even before the forms are erected, the level of the excavations should be checked, using a long, straight piece of 2 by 4 and a mason's level.

Where stepped footings are involved, such as those shown in Figure 7, each step, such as from *a* to *b*, *c* to *d*, and *e* to *f*, must be carefully checked to make sure it is perfectly level. The risers, such as *bc* and *de*, must be exactly vertical. If any appreciable slope or lack of levelness occurs in such footings, they might slip, move, or break when the full foundation load is applied to them.

DESIGN OF FOOTINGS

In most cities and towns the building codes specify the kinds and dimensions of footings to use for all types of structures. Readers are advised to check such codes in their localities. Where no codes exist, such as in rural areas, the following design procedures may be employed:

General Design. For ordinary types of residences and for other small one- and two-story buildings to be erected on soils of average load-carrying capacity, concrete footings can generally be designed following the proportions shown in the *A* part of Figure 2. In that proportion, the letter *W* means the thickness of the foundation. Suppose, for example, that a foundation is to be 10 inches thick. Then, *W* = 10 inches. The width of the footing would be 2*W* (two times *W*), or 20 inches. The thickness would be 10 inches. The *W*/2 projections would each be 10 ÷ 2, or 5 inches wide.

When footings such as those shown in Figure 3 are to be used for ordinary small buildings and on soil of average load-carrying capacity, the *W* proportions indicated can be used for design purposes.

Special Design. When heavier-than-usual houses and other buildings are involved, special design calculations must be made which are somewhat more complicated but not difficult. The actual weights of all walls, floors, and roof must be determined so that the exact load that 1 lineal foot of the footings must support will be known.

Example. Figure 9 shows a rough section view of a two-story house. Only the principal structural members and approximate dimensions are included. The foundation can be assumed to be made of concrete, the

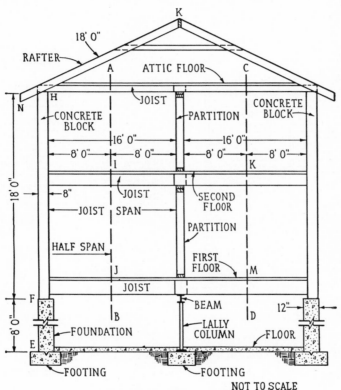

FIGURE 9. Cross section of house.

exterior walls of concrete block, and all other details of wood. The soil can be assumed to be composed of firm clay. The problem is to calculate the total load that 1 lineal foot of the footing, shown at *E*, will have to support and to design the footing.

It should be realized that the necessary calculations include only 1 lineal foot of the foundation, only 1 lineal foot of the exterior wall, and only 1 lineal foot of each of the three floors. The *A* part of Figure 10 will be helpful in visualizing what this means. The black square at *G* indicates 1 lineal foot of the wall. The dark outline from *G* to *H* indicates that part of each floor, 1 foot wide, which must be considered. The part of the floor from *H* to *J* is supported by the partition *EF*. In addition to the foregoing, a 1-foot-wide section of half the roof, marked *NK* in Figure 9, has to be included.

Floors and roofs are subject to *dead loads* (weight of the building materials) and to *live loads* (weight of furniture, snow, wind pressure,

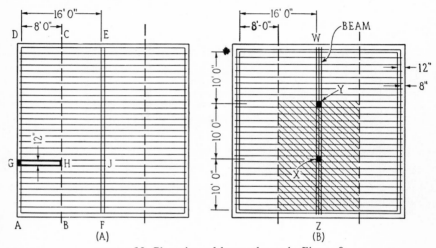

FIGURE 10. Plan view of house shown in Figure 9.

equipment, and occupants). For the sake of this example, the necessary dead and live loads are assumed. NOTE: All building codes specify what dead and live loads shall be.

ROOF. Assume a combined dead and live load of 40 pounds per square foot. As shown in Figure 9, the rafters are each 18′ 0″ long. The total roof load for a section 1 foot wide and 18′ 0″ long is $18 \times 40 = 720$ pounds.

ATTIC FLOOR. Assume a combined dead and live load of 60 pounds per square foot. For a section of this floor 1 foot wide and 8′ 0″ long (see Figures 9 and 10), the total load is $60 \times 8 = 480$ pounds.

SECOND FLOOR. Assume a dead and live load of 70 pounds per square foot. For a section of this floor 1 foot wide and 8′ 0″ long, the total load is $70 \times 8 = 560$ pounds.

FIRST FLOOR. Same as second floor.

WALL. The concrete blocks in the exterior wall weigh 60 pounds per cubic foot. One lineal foot of the wall is 8", or ⅔ foot thick, 1' 0" long, and 18' 0" high. Multiplying ⅔ × 1 × 18 = 12 cubic feet. Then, 60 × 12 = 720 pounds.

FOUNDATION. One lineal foot of the foundation is 1' 0" thick, 1' 0" long, and 8' 0" high. Multiplying 1 × 1 × 8 = 8 cubic feet. Then, 150 × 8 = 1,200 pounds. Assume the footing weighs 300 pounds.

The total load, per lineal foot, is as follows:

Roof load	720 pounds
Attic-floor load	480 pounds
Second-floor load	560 pounds
First-floor load	560 pounds
Wall load	720 pounds
Foundation load	1,500 pounds
Total load	4,540 pounds

The bottom area for the necessary footing is calculated as follows:

The firm clay soil can support a load of 4,000 pounds per square foot. Thus, the load of 4,540 pounds per lineal foot hardly needs a footing. However, most designers and builders would follow the *W* proportions shown in the *A* part of Figure 2.

To design the footings for the house shown in Plates I through VI, exactly the same procedure should be followed, except that the house has only one story. The 20-inch-wide footing shown is in excess of the needs. However, it was decided upon as a means of making absolutely sure that cracking and settlement could never occur.

Design of Column Footing. There is no general way of designing a column footing. This is because columns carry concentrated loads which vary considerably with every house or other building.

Example. Figure 9 shows that Lally columns are to be used to support the beam under the interior partition of the house. The *B* part of Figure 10 shows the same beam which is supported by the columns at *X* and *Y* and by the foundation at *W* and *Z*. The problem is to calculate the size of footing required under column *X*. This column supports the load from one end of the beam running from X to Z and one end of the beam running from X to Y. This means that the total floor and partition loads from the attic, second, and first floors must be calculated between lines AB and CD, as shown in Figure 9, and between support Z and column Y, as shown in Figure 10.

As shown in Figure 9, the floor spans from A to C, I to K, and J to M. All these spans are 16′ 0″ wide. The distance between Z and Y in Figure 10 is 20′ 0″. The area of each floor to be dealt with is 16 feet times 20 feet, or 320 square feet. The attic-floor load was given as 60 pounds per square foot in the previous example. The total attic-floor load is 320 × 60, or 19,200 pounds. The second and first floors each have an area of 320 square feet and a floor load of 70 pounds per square foot. The load from each floor is therefore 320 × 70, or 22,400 pounds.

The partitions, directly over the beam, can be assigned an approximate weight of 110 pounds per lineal foot. The partitions on the first and second floors are each 20 feet long. Their combined length is 40 feet. Then, 110 × 40 = 4,400 pounds.

The total load which the beam must support between XZ and XY is as follows:

Attic-floor load	19,200 pounds
Second-floor load	22,400 pounds
First-floor load	22,400 pounds
Partition load	4,400 pounds
Total load	68,400 pounds

The beam between X and Z, as shown in Figure 10, supports half of 68,400 pounds, or 34,200 pounds. The beam between X and Y supports a like load. The beam between X and Z exerts half of its load on the column at X, or 17,100 pounds. The beam between X and Y exerts a like load on column X. Thus, column X must support a load of 34,200 pounds. For simplicity in calculation, the weights of the beams and columns are ignored.

If the soil consists of firm clay, it can support a load of 4,000 pounds per square foot. In such case, the bottom area of the footing must be 34,200 ÷ 4,000, or approximately 8.55 square feet. For ease in calculation, this can be called an even 9 square feet. This means that the bottom of the footing must be 3′ 0″ square.

Following the proportions for the footing shown at *A* in Figure 2, the value of $2W$ is 3′ 0″, or 36 inches. The value of *W* is one-half of 36 inches, or 18 inches. Thus the footings must be 18 inches thick and 36 inches square. When a footing is this wide, reinforcement should be used.

Following the proportions of the footing shown at *F* in Figure 3, the value of *W* is governed by the base plate of the Lally column. Such plates are generally 12 inches square. Then with *W* equal to 12 inches, the projection will be 6 inches; the step, 6 inches; the total thickness, 24 inches; and the bottom width, 36 inches.

LAYOUT

In order to explain and illustrate the method of laying out a building, the following example for a one-story, basementless house is presented. The procedure employed does not require the use of surveying instruments. For other and more elaborate layouts, as for houses or buildings located on sloping sites, for houses of several levels, and where basements are required, a professional surveyor should be employed to establish boundaries and levels.

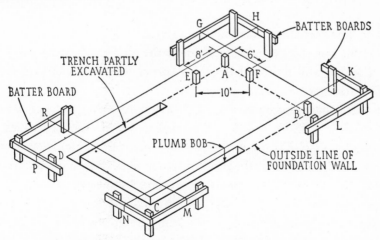

FIGURE 11. Right-triangle method of laying out boundary lines.

Laying Out the House. First, a base line is established, marking out one end or one side of the proposed house (see line *AB* in Figure 11).

Stakes are set at *A* and *B* on this line, locating two corners. The distance between the stakes represents the over-all width or length of the house. In the top of stake *A,* a nail should be driven near its center. This nail accurately locates the corner.

To lay off the corner (using what is called the right-triangle method), drive a stake at *F,* as shown in Figure 11, so that it is along *AB* and 6 feet from stake *A.* A nail should be driven in the top of this stake exactly 6 feet from the nail in the stake at *A.* Stake *E* should be driven so that its center line will be exactly 8 feet from stake *A* and exactly 10 feet from

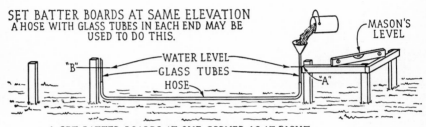

SET BATTER BOARDS AT SAME ELEVATION
A HOSE WITH GLASS TUBES IN EACH END MAY BE
USED TO DO THIS.

MASON'S LEVEL

"B" — WATER LEVEL — GLASS TUBES — HOSE — "A"

1. SET BATTER BOARDS AT ONE CORNER AS AT RIGHT.
2. PLACE HOSE AS SHOWN.
3. FILL WITH WATER UNTIL WATER LEVEL IS AT TOP OF BATTER BOARD "A".
4. MARK WATER LEVEL AT OPPOSITE END "B" AND SET BOARD TO MARK.

FIGURE 12. Setting batter boards.

stake *F.* The corner representing the angle *EAF* is a right triangle corner (90°); the line *AE* extended to *D* forms the second boundary line of the house, and *D* will represent the third corner. Other corners are located in a like manner.

After this has been done, strings (heavy cords) are stretched over the corner stakes *A, B, C,* and *D* and carried to outside supports called *batter boards* (*G-H-K-L-M-N-P-R*). The tops of the horizontal pieces (batters) should be set at the proposed floor level. The building lines may be projected from the strings to the ground by means of a plumb bob suspended as shown in Figure 11. When the batter boards have been set (see Figure 12) and the strings indicating the layout of the house transferred to them, the corner stakes *A, B, C,* and *D* and stakes *E* and *F* can be removed so that the excavation can be made. Nails should be driven in the batters where the strings are fastened, so that, if the strings are broken or removed, they can be replaced accurately.

Setting Batter Boards. Figure 12 indicates how to set all batter boards at the same elevation.

EXCAVATION

There are several important details to consider before excavation work is started and while it is in process. All such details, if properly taken care of, contribute in no uncertain manner to the construction of good footings.

Frost Line. In all localities where severe winter freezing is likely, the excavation should be extended well below the line of maximum frost penetration. The amount of penetration varies according to locality, going down to 4 feet or more in some areas. Local weather-bureau stations can supply such data.

When soils freeze, they expand considerably. Then, as they thaw, an equal amount of contraction occurs. This is often known as *heaving*. If footings are not placed well below the maximum frost line, the heaving can crack them, injure foundations, and cause severe damage to other parts of buildings.

Excavating Procedure. When footings, such as shown at *A* and *B* in Figure 2, at *E* and *F* in Figure 3, at *K, L, M,* and *N* in Figure 6, and in Figure 7, are to be placed, the trench (see Figure 11) should be amply wide to allow for the footing width. It should be kept in mind that the outside edge of such footings will be several inches beyond the outside line of the foundation, as shown in Figure 11.

In firm soils the sides of the trenches, for the type of house considered here need not have much slope. However, if the soils are loose, the slope must be increased to the extent where sliding or caving will not occur. Ordinarily, the trench is excavated 2 or more feet wider than the width of the footing as a means of allowing ample room for installation of forms. The depth of the trench depends upon frost depth, whether or not a crawl space is required, and the height of the foundation. In cases where footings like those shown at *H* in Figure 4 and at *I* and *J* in Figure 5 are required, the depth of excavation is not much of a problem. In any event, once the depth of the footing has been decided upon, a plumb bob, as shown in Figure 11, can be used to gage the depth.

The bottom of the trench should be in *firm* and *undisturbed* soil. Thus, care should be exercised not to excavate too deeply so that backfilling is required.

The bottom of the trench must also be level. This can be determined by the use of a straight 2 by 4, which is at least 8 or 10 feet long, and a mason's level. The 2 by 4 should be placed on edge at various points along the length of the trench and the level laid on it.

The bottom of the trench should be at the same elevation (distance below horizontal parts of batter boards) at all points. This can be checked by suspending a plumb bob from the strings as explained under Layout in this chapter.

FORMWORK

Formwork is recommended for all footings except where exceptionally lightweight buildings are concerned. The forms are an aid in making the footings square, level, and of the proper shape and size.

Figure 13 shows the construction of typical forms for footings. It is recommended that 2-inch lumber be used as a means of producing forms which are solid and true. Note how the vertical side forms are held in place by stakes which should be spaced at intervals of about 4 feet. The stakes in turn can be held upright by other stakes and braces. Spacers should be used at about 4-foot intervals to prevent the side forms from leaning.

As the side forms are set into position and just before the concrete is placed, a mason's level should be placed at intervals along them and across, from one to the other, to make sure that they are both level and both at the same level.

The use of two-headed nails will be an aid when the forms are being removed. When nails can be pulled without strain, there is less chance of injury to the new concrete.

In all the footing illustrations, such as at *A* in Figure 2 and at *E* and

F in Figure 3, grooves are shown in the top surface of the footings. Such grooves tend to prevent any possible sliding of the foundations. A typical groove form is shown in Figure 13. Generally, such grooves are triangular, with the sides of the triangle about 3 inches long.

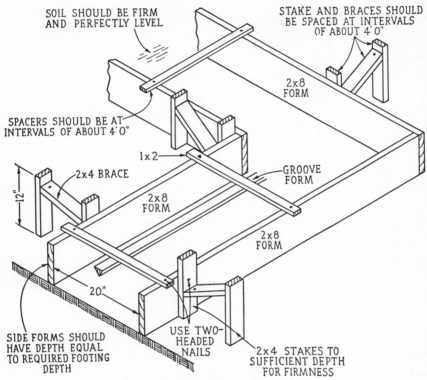

FIGURE 13. Typical forms for footings.

Figure 14 shows typical forms for a column and a curb. Such footings should be installed with all the care explained in the foregoing discussions.

MIX AND MIXING

Footings, because they are not subject to wear or weather, can be made using a 7-gallon mix, as shown in Table 1 of Chapter 4, Concrete. However, if a stiffer mixture is desired, the water can be reduced to 5½ gallons per bag of cement. The proportion of sand and stone can be

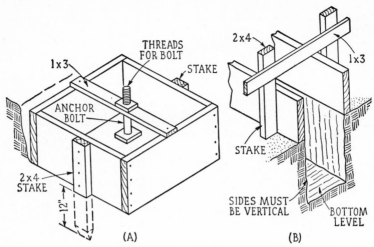

FIGURE 14. Typical column footing and curb forms.

varied to make a workable or mushy mix, but either one or the other amount of water should be used without change. When power mixers are used, the mixing should be continued for at least 2 minutes. When mixing is done by hand, the mixture should be turned over and over until it is all one color and workable.

PLACING CONCRETE

Concrete should be placed as nearly as possible or practical in its final position. To avoid segregation, it should not be placed in large quantities at a given point and allowed to run or be worked over a long distance. As concrete is placed, it should be spaded, using a shovel or regular spading tool, just enough to compact it and to avoid any possibility of pockets or voids, especially next to the forms.

If delay in the placing of concrete occurs, and the concrete starts to stiffen, it may be safely used if its workability is restored by remixing. However, no more water should be added.

FINISHING

After the concrete is placed in the forms, a short piece of 2 by 4 can be used, across the two side forms, to level the surface. The 2 by 4 can be zigzagged back and forth until a smooth, even surface is obtained. When groove forms are used, care should be exercised to make sure that the concrete is well packed around and underneath them.

FORM REMOVAL

During warm weather, the forms can be safely removed after 2 or 3 days. During cold weather, the forms should not be removed for at least a week.

CHAPTER SIX

Concrete Foundations

Every well-designed house or other building is actually planned by starting below the surface of the soil. In fable, verse, and Scripture man has been warned to build on firm foundations. Part of such warning was heeded in the suggestions concerning footings, set forth in Chapter 5. An equally important part of the warning has to do with foundations which are supported by footings.

Because foundations are concealed from view after a building has been completed, there is some tendency to overlook their great importance. Such a practice should be avoided because the stability and permanence of any building are completely dependent upon their proper design and construction. A poorly made foundation often accounts for unequal settlement, cracks, tilting, sagging floors, stuck doors and windows, and many other most undesirable aspects, which may result in unsafe conditions. Since foundations are the main support for so much of any building, they should be given careful attention and consideration.

Good foundations are not difficult to plan and build. With the proper attention to planning and with the exercise of appropriate care in building, they can be constructed to meet every need. The purpose of this chapter is to explain the necessary principles involved.

PURPOSE OF FOUNDATIONS

The purpose of foundations can best be explained by a careful consideration of the important functions they serve. In the following discussion, only such simple foundations as are commonly found in houses and other small buildings are involved:

Foundations Provide Support. Previous chapters have repeatedly emphasized the point that concrete must be well supported before it, in turn, can safely and surely support loads. This same fact applies to foundations and to all other structural work in houses and other buildings.

The question of support must be considered constantly because every part of any building depends upon some other part or parts of the same building for its support. For example (as explained in Chapter 5, Footings), the roof of a two-story house is usually supported by two or more of the exterior walls. Portions of the attic, second, and first floors are also supported by these walls. These walls are, in turn, supported by the foundations. Thus, foundations are the means of supporting much of the load from the structural work (both dead and live loads) above them.

The load must be supported in such a manner as to provide equal and positive strength at all points under the exterior walls so that there can be no chance of seesaw (up-and-down) action. In other words, because of unavoidable and unequal loading on various floors, etc., the loads at various points along the foundations may vary. The foundations must be able to absorb such varying loads without allowing any settlement or cracking.

Foundations Give Frost Protection. Foundations, properly combined with footings, prevent damage from heaving, so far as up-and-down movement is concerned. But there is also a heaving action along horizontal lines which foundations must resist to prevent damage to a building. Such horizontal heaving tends to exert a force which, without good foundations, can twist the wood frame and other structural details.

Foundations Provide Basements. If basements are desired, the foundations hold back the soil and, if properly constructed and waterproofed, keep such basements dry.

Foundations Guard against Termites. In many localities, termites are apt to eat and seriously damage any wood structural detail which comes in contact with the soil. Masonry foundations keep wooden parts away from the soil where termites cannot easily reach them. However, it should be pointed out that metal termite shields and especially treated lumber, as specified by most city building codes, should also be used.

Foundations Resist Earthquake Shocks. Because of the nature of earthquakes, they tend to pull structural details apart. The rolling or whip-cracking action causes opposing stresses which tear and split materials. Good concrete foundations, because they form one integral mass, roll with earthquake movements and resist or tend to eliminate stresses which would cause damage to plaster and other structural parts above them. This is one reason why foundations, as explained in subsequent pages, should be placed all at the same time.

KINDS OF FOUNDATIONS

Figure 1 shows a picturelike illustration of the various kinds of concrete foundations which are usually constructed for average one- and

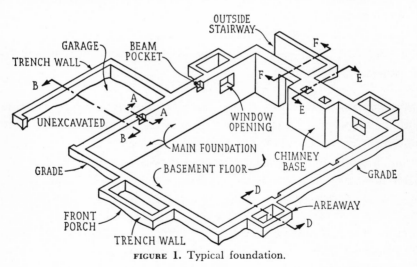

FIGURE 1. Typical foundation.

two-story houses. It is also an example of the integral foundation mentioned previously. When all parts of such foundations are placed at the same time, they constitute one piece. This is the desirable quality which helps to prevent earthquake damage. The illustration shows the main foundation, a retaining wall for an exterior basement stairway, retaining walls for areaways, and a porch foundation.

Main Foundation. In Figure 1, there is a cutting line, *A-A,* shown in connection with the main foundation. In Figure 2, at *A-A,* a section view of that foundation is illustrated. Note the *beam pocket* shown in both illustrations. Such pockets support beam ends such as shown at *W* and *Z* in Figure 10 of Chapter 5.

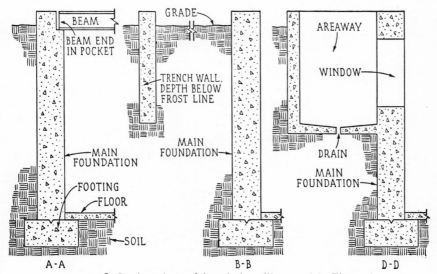

FIGURE 2. Section views of foundations illustrated in Figure 1.

Garage Trench Walls. In Figure 1, there is a cutting line, *B-B,* shown in connection with the garage area. In Figure 2, at *B-B,* a section view of that same trench wall, along with the main foundation, is shown so as to indicate the relationship. Small foundations used for lightweight loads and which require no footings are known as *trench walls.*

Areaway Retaining Wall. In Figure 1, there is a cutting line, *D-D,* shown in connection with a basement window. An areaway, or opening around basement windows, is built so that light and air can enter windows below grade line. In Figure 2, at *D-D,* a section of that same retaining wall, along with the main foundation, is shown so that the relationship between the two can be visualized.

Basement-stairway Retaining Wall. In Figure 1, there is a cutting line, *F-F*, shown in connection with the basement exterior stairway area. In Figure 3, at *F-F,* a section view of that same retaining wall is shown, along with the opening in the main foundation for the door. Retaining walls of this kind do not have to support loads from above them. Their only function is to hold back the soil. Thus, footings are not generally

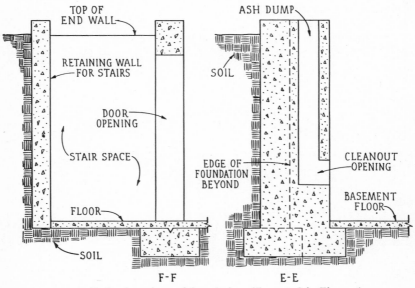

FIGURE 3. Section views of foundations illustrated in Figure 1.

necessary, especially if they extend below frost level in a sandy or rocky soil.

Chimney Foundation. In Figure 1, there is a cutting line, *E-E*, shown in connection with a fireplace chimney. In Figure 3, at *E-E,* a section view of that same chimney base is shown as part of the main foundation. Note that the section view shows an ash dump and a cleanout opening. Such a chimney base should be an integral part of the foundation.

Porch Foundation. The porch foundation shown in Figure 1 is so nearly like the trench walls for the garage that no special illustration is required.

Foundations and Piers. The aim of Figure 4 is to illustrate general procedure relative to foundations and piers. In many instances, where one-story houses are to have crawl spaces between the underside of the joists and the surface of the soil, a combination of foundations and piers are used, somewhat as shown in the *X* and *Y* parts of Figure 4. The methods of using foundations and piers vary widely between localities, building codes, and builders. There is no standard design which applies to all parts of the country. However, the piers are most generally used in

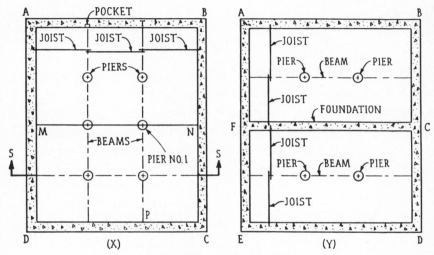

FIGURE 4. Foundation and piers.

warm climates. The use of such foundations and piers is subject to local building codes and may be prohibited in many Northern and Eastern regions.

In the *X* part of the illustration, a standard type of foundation and footing may be used under all exterior walls, as shown by *ABCD*. Or a type of foundation similar to a trench wall may be employed if the soil is composed of sand and gravel. The piers are also made of concrete. The beams may be built up of two or three joists, or they may be steel I-beams.

The *Y* part of the illustration shows how an added length of standard foundation is sometimes employed. Fewer piers are then required.

Figure 5 shows a section along the cutting line marked *S-S* in Figure 4. The piers may have any one of several types of bases, such as shown in the

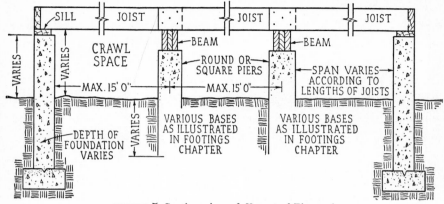

FIGURE 5. Section view of *X* part of Figure 4.

B and *C* parts of Figure 2 and the *G* part of Figure 4, all of which are discussed in Chapter 5, Concrete Footings. Many of the dimensions, such as height of foundation above grade, height of crawl space, depth of foundation, and depth of pier bases, vary according to localities, especially between warm and cold climates.

Pilasters. In some cases, especially where long and heavily loaded beams are supported at their ends by the foundations, ordinary pockets, as shown in Figure 1, cannot be used because the foundation thickness is not great enough to support the load amply. For example, the area *ABEF* in Figure 6 may not be sufficient surface, in terms of square inches, to distribute a beam-end load to the point where the per-square-inch load does not exceed the compressive strength of concrete. In such cases, a *pilaster,* as shown in Figure 6, must be placed as an integral part of the foundation. The area *ABCD* increases the bearing area upon which the beam end rests so that its load is distributed within the per-square-inch compressive-strength limitations of the concrete. Pilasters also add sta-

FIGURE 6. Typical pilaster.

bility to foundations at points where heavy loads tend to act in other than a straight-down direction.

PLANNING FOUNDATIONS

Practically all cities and towns have building codes which set forth definite and rigid specifications regarding the planning of foundations. Therefore, masons are urged to study their local code before doing any planning work of their own. In cases where codes do not exist, such as in rural areas, the following suggestions can be employed:

General Planning. The following general rules apply to houses and other small buildings:

1. One-story houses:
   a. Wood frame without basement. Make the foundations at least 6 inches thick.
   b. Wood frame with basement. Make the foundations at least 8 inches thick, or at least 10 inches thick if they are longer than 20 feet.
   c. Masonry exterior walls. Make the foundations at least as thick as the walls they support.
   d. Brick or stone veneer with wood frame. Make the foundations at least 8 inches thick if veneer does not extend more than 1¼ inches beyond the edges of the foundations. Otherwise increase the thickness of the foundation to the extent that the veneer does not extend more than 1¼ inches beyond the edge of the foundations.
2. Two-story houses with basements:
   a. Wood frame. Make the foundations 10 inches thick if they do not extend more than 7′ 0″ below grade line. Make them 12 inches thick if they extend below 7′ 0″.
   b. Masonry walls. Make the foundations at least the same thickness as the walls they support if the foundation does not extend more than 7′ 0″ below grade line. If the foundation extends more than 7′ 0″ below grade line, make it 12 inches thick.

c. Brick or stone veneer with wood frame. Make the foundations at least 10 inches thick if they do not extend more than 7′ 0″ below grade line. Make them 12 inches thick if they go below 7′ 0″.

3. Two- and three-story apartment buildings with basements:
   a. Wood frame (see item 2*c*).
   b. Masonry walls. For a two-story building make the foundations at least as thick as the walls they support; for a three-story building make them 12 inches thick.

4. One-story store building:
   a. Wood frame without basement. Make the foundations at least 8 inches thick, or at least 10 inches thick if they are longer than 25 feet.
   b. Wood frame with basement. Make the foundations at least 10 inches thick, or at least 12 inches thick if they are longer than 25 feet.

5. Two- and three-story store buildings with basements:
   a. Masonry walls (see item 3*b*).

6. One-story dairy barns:
   a. Wood frame. Make the foundations at least 8 inches thick if they are not over 25 feet long and at least 10 inches thick if they are longer than 25 feet.
   b. Masonry side walls. Make the foundations at least as thick as the walls they support and at least 12 inches thick if they are longer than 25 feet.

7. Dairy barns with haymows:
   a. Wood frame. Make the foundations at least 10 inches thick if they are not over 25 feet long and 12 inches thick if they are longer than 25 feet.
   b. Masonry walls. Make the foundations at least as thick as the walls they support and at least 12 inches thick if they are more than 25 feet long.

8. One-story-house garages without basements:
   a. Wood frame. Make the foundation at least 8 inches thick.
   b. Masonry walls. Make the foundations at least as thick as the walls they support but not less than 8 inches.

9. Miscellaneous small farm structures:
   a. Wood frame. Make the foundations at least 8 inches thick, or at least 10 inches thick if over 25 feet long.
   b. Masonry walls. Make the foundations at least as thick as the walls they support but not less than 8 inches, or 10 inches if they are more than 25 feet long.
10. Residence porch foundations (see Figure 1):
    Make the foundations not less than 5 inches thick.
11. Residence areaway walls (see Figure 1):
    Make the walls at least 6 inches thick.
12. Residence exterior-stair walls (see Figure 1):
    Make such walls at least 8 inches thick if not more than 10 feet long.
13. Chimney bases (see Figure 1):
    When a chimney base is part of a foundation, its thickness should conform to the horizontal area of the fireplace and be that much thicker than the foundation.

Foundations and Piers. For one-story houses the foundations and piers, as illustrated in Figures 4 and 5, may conform to the following:
1. One-story houses:
   a. Wood frame. Make the foundations at least 8 inches thick, or at least 10 inches thick if they are longer than 25 feet.
   b. Make the piers at least 12 inches in diameter, or square when the distance between them and between them and the foundations is not more than 15 feet.

Pilasters. In houses and other small buildings, the planning of pilasters generally depends on the thickness of the foundations of which they are to be parts. For all general purposes, pilasters should be planned as shown in Figure 6. The added thickness which the pilasters provide should be at least equal to the thickness of the foundation. The width, such as *CD*, should be at least 12 inches.

When pilasters are used as a means of adding stability to long foundations, they may be used at about 25-foot intervals.

For buildings larger than ordinary houses, small stores, etc., a structural engineer should be consulted about foundation and pilaster thicknesses.

The same is true in cases where foundations must be more than 10 feet high, from the footing to their top.

Layout and Excavation. In all instances where footings are employed, the layout and excavation will have been completed by the time foundation work is to be started (see Layout in Chapter 5, Concrete Footings).

The layout, excavation, and installation procedures for piers, such as shown in Figures 4 and 5, can be accomplished as per the following typical example in which the type of pier shown at *G* of Figure 4 in Chapter 5, Concrete Footings, is assumed.

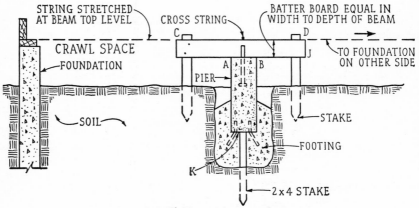

FIGURE 7. How to lay out piers.

Once the foundations and sills, as shown in Figure 5, have been placed, strings can be stretched across the site, from sill to sill, in such a manner that they cross at the pier locations. For example, as shown in the *X* part of Figure 4, strings can be stretched from *M* to *N* and from *P* to *T*. Where they cross indicates the position of pier 1. A plumb bob can be dropped from the point of intersection of the strings to mark the exact position of the pier.

Excavate a circular hole whose diameter is about twice the diameter of the pier, if the pier is round, or about 2 feet square if the pier is square. The sides of the excavation should be as nearly vertical as the soil will allow. If the soil is dry sand, a rough wood form will have to be provided.

The batter board, as shown in Figure 7, should have a width, *DJ*, which is equal to the depth of the beam (see Figure 5). Stakes should be used to keep the top edge, *CD*, of the batter board level with one of the strings.

A 2 by 4 stake should then be driven part way in and the bottom of the precast pier placed on it as shown at *K* in Figure 7. The stake and the pier can be pounded down until the top, *AB*, of the pier is level with the underedge of the batter board. The anchor bolt in the pier can be tied to the batter board with wire until the concrete has been placed and has hardened.

FOUNDATION FORMWORK

Foundation forms should be carefully built and strongly braced to ensure foundations that are true to line and grade. The importance of this can be illustrated and explained by the following example:

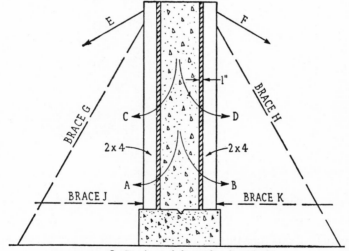

FIGURE 8. Section of form for foundation.

Suppose that part of a foundation is 12 inches thick, 8 feet high, and 15 feet long. The cubic content of such a foundation would be $1 \times 8 \times 15$, or 120 cubic feet. Concrete weighs 150 pounds per cubic foot. Therefore, the total weight would be 150×120, or 18,000 pounds. That is a weight of 9 tons!

Figure 8 shows a section view of the foundation being discussed, along with forms constructed of 2 by 4s and 1-inch plywood. The weight of the concrete exerts pressure in several ways. Arrows A and B show the direction of pressure near the bottom of the forms. Arrows C and D show the direction of pressure at about mid-height of the forms. Arrows E and F show the direction of pressure which tends to make the forms fall over. To overcome such pressures, the forms must be strong and braced somewhat as shown by braces J, K, G, and H.

Even if only 1 lineal foot of such a foundation is considered, the weight would be 1,200 pounds. Just that much weight is a substantial pressure which only strong, well-braced forms can safely withstand until the concrete hardens and develops strength of its own.

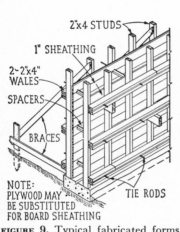

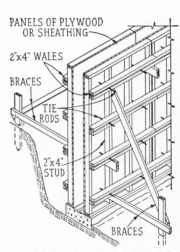

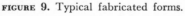

FIGURE 9. Typical fabricated forms.   **FIGURE 10.** Typical prefabricated forms.

Common Types of Formwork. Two types of formwork are commonly used for foundations. The first type (see Figure 9) is the more conventional method of erecting forms, using wood sheathing or plywood, studs, wales, and bracing. Usually this method is used for small or infrequent projects. The second type, as shown in Figure 10, employs the use of prefabricated panels that are assembled on the job site. These panels, or so-called *unit forms,* are built in various sizes so that they can be readily removed from one job and reassembled on subsequent work.

For fabricated forms, as shown in Figure 9, 1-inch plywood is commonly used for sheathing. When the plywood is supported by 2 by 4 studs at 24-inch intervals, it will withstand pressure without bulging or failing when concrete, up to a depth of 3 feet, is placed in the forms. When concrete must be greater in depth, the studs should be placed at intervals closer than 24 inches and the sheathing must be thicker.

Prefabricated forms, as shown in Figure 10, are usually made of wood or metal. Wood panels are made of sheathing and studs, plywood backed up by sheathing and studs, or plywood backed only by studs. Metal panels are usually made from sheet steel.

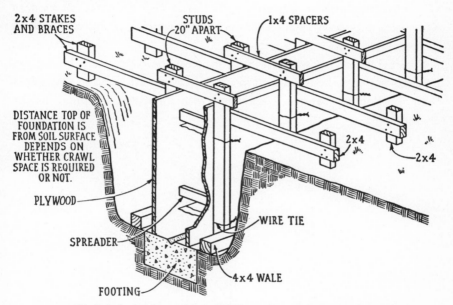

FIGURE 11. Fabricated forms for foundation where no basement is required.

The desired smoothness of a foundation surface often determines the choice of form materials. When a particularly smooth finish is desired, plywood or form-lining paper can be used. Joints between boards or panels should be tight to prevent loss of cement paste. When paste is lost through cracks, the concrete is weakened and is apt to be honeycombed.

Wire ties are used to keep the two form faces from spreading apart. Soft black annealed iron wire is generally used for the wire. It is strong in tension (cannot be pulled apart easily) and will withstand the severe twisting that is necessary. The wires are cut flush with the foundation surface when the forms are removed.

In many instances, tie rods are used instead of wire ties. The rods are easier to install, and they need not be as closely spaced as wire. Several kinds are available in all shapes and sizes.

When spreader attachments are not part of the ties, wood spacers are commonly used to keep the inside surfaces of the forms the proper distances apart. These are cut to a length equal to the desired wall thickness and are placed between the forms before the ties are tightened. The spacers are removed as the placed concrete reaches their level.

To prevent concrete from sticking to the forms, crude oil, soft soap, or whitewash may be applied on the form faces which come in contact with the concrete.

Figure 11 shows how fabricated forms can be constructed for a basementless one-story house of the type discussed under Layout in Chapter 5, Concrete Footings, and in connection with Figures 4, 5, and 7 in this chapter.

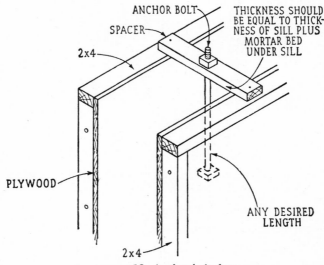

FIGURE 12. Anchor-bolt form.

Anchor-bolt Form. There is some tendency simply to stick anchor bolts into concrete without the use of forms or other means of making sure that the bolts are absolutely vertical. Such a practice should be avoided. When anchor bolts are required, as is generally the case, they should be hung in position by the use of a simple form, such as shown in Figure 12. The underside of the spacer should be at the level of the foundation top.

Insert for Prefabricated Forms. There are instances where the prefabricated-form units cannot be planned to fill the needs exactly. In such cases the use of inserts, as shown in Figure 13, can be resorted to. The inserts can be made, using boards and 3 by 3s or 2 by 4s. The inserts can be made any width and height.

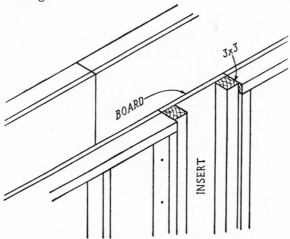

FIGURE 13. Insert used with prefabricated form.

Window-opening Forms. Where windows are required in foundations, as shown in Figure 1, the use of a form such as shown in Figure 14 is necessary. Such forms should fit tightly against the main forms to prevent leakage of cement paste. The key strip can, if so desired, be made of triangular pieces of wood which will be keyed to the hardened concrete and serve as a nailing strip for the window frames.

Door Openings. Where doors are required in foundations, as indicated in Figure 1, the necessary forms are similar to those for window openings except that the sides, marked X in Figure 14, are the actual door frames with keys attached to them. Cross bracing should be used at 2-foot intervals to resist the pressure of the concrete weight.

Beam Pockets. Where beam pockets are required, as indicated in Figure 1, the necessary forms can be made similar to the ones suggested in Figure 15. The pieces marked X hold the pocket form in place so that it cannot sag or push outward.

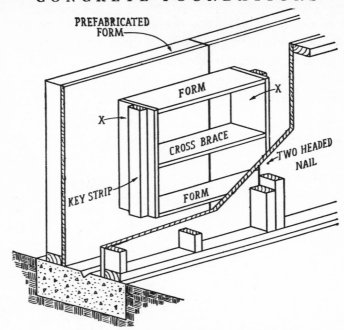

FIGURE 14. Form for window opening.

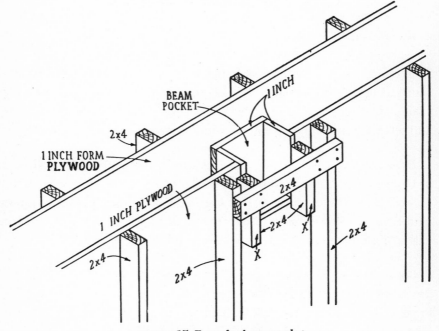

FIGURE 15. Form for beam pocket.

Utility Connections. Openings in foundations to accommodate the passage of utility connections, such as water and gas pipes, are formed by metal or wood sleeves, somewhat larger in diameter than the pipes, and attached to the forms. After the concrete hardens and the forms and sleeves are removed, the pipes can be set in place. To make the foundation watertight, the space between the pipes and the concrete should be tightly caulked with oakum or filled with stiff mortar.

Two-headed Nails. It is suggested that two-headed nails be used when constructing formwork. Such nails can be pulled out easily and without pounding, which might damage new concrete or ruin the form lumber.

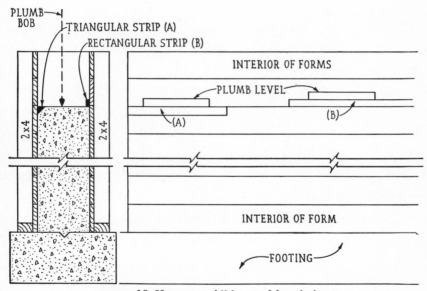

FIGURE 16. How to establish top of foundation.

Foundation Height. In most instances where prefabricated or unit forms are used, the top of the foundation will be lower than the top of the forms, as shown in Figure 16. In order that the top of a foundation will be at the correct height at all points along the forms, the interior of the forms must be marked to indicate where the top of the concrete should be. Without surveying instruments, this can be accomplished as follows:

Stretch strings between the batter boards, set up as explained in Chapter 5, Concrete Footings. From these strings, a plumb bob can be suspended to indicate where the top of the foundation should be. Then nail triangular or rectangular strips, as shown in Figure 16, to the inside faces of the forms. The triangular strips will bevel the corner of the foundation, while the rectangular strips will make the corners square.

Start at any corner of the forms and lightly nail the strips in place. A plumb level should be used, as shown in Figure 16, to make sure that the strips are level. Nail strips all the way around the interior faces of the forms until the strips meet the first ones at the starting corner. If the last strips do not exactly coincide with the first strips placed, some inaccuracy exists, and the level, all the way around, must be rechecked using the same method. When all strips are finally at the same level, they can be securely nailed to the forms.

The foregoing method is not as exact as when surveying instruments are used. However, for most purposes it is accurate enough if carefully done.

MIX AND MIXING

Table 1 shows the recommended mixes to use in terms of gravel or crushed-stone sizes. These mixes conform to the specifications given in Chapter 4, Concrete.

TABLE 1. Recommended mixes for foundations

| Maximum size of aggregate, inches | Mixes | | | |
| | | | Aggregates | |
	Mixing water* per bag of cement, U.S. gallons	Portland cement, bags	Sand, cubic feet	Gravel or crushed stone, cubic feet
$\frac{3}{4}$	5	1	$2\frac{1}{2}$	$2\frac{3}{4}$
1	5	1	$2\frac{1}{4}$	3
$1\frac{1}{2}$	5	1	$2\frac{1}{4}$	$3\frac{1}{2}$

* Based on 6 gallons of water per bag of cement including water contained in damp sand.

The concrete should be mixed thoroughly until it is uniform in appearance and all ingredients are evenly and well distributed through the mixture. Machine mixing should continue for at least 2 minutes.

The concrete should be placed in the forms promptly after mixing. When any delay in placing occurs and the concrete has become somewhat stiff, it may still be used if its workability can be restored by remixing *without* additional water.

PLACING

Special care should be taken in placing the concrete in the forms. It should be placed in layers 6 to 12 inches in depth and deposited in the forms at close intervals not exceeding 6 feet. Such practice avoids separation of materials and assures uniform mixture throughout a foundation. As concrete is placed, it should be spaded to make sure that no voids occur, especially next to the forms. Spading also settles the mixture into a dense mass which is strong and watertight with surfaces free from defects.

Foundations should be placed in one continuous operation so that they constitute one integral mass.

WATERPROOFING

Unless foundations are erected in sections of the country where dry climates are the rule, or where the subsoil is exceptionally well-drained and dry, some means of waterproofing should be adopted.

Where rain and poorly drained subsoil are the rule, waterproofing, such as shown in Figure 17, is recommended. This does not necessarily mean that concrete will leak or that previously explained aspects of watertight concrete are not true. Instead, it is simply an added means of making sure that basements will be dry.

The joint between the floor and the foundation should be filled with asphaltum. The outside surface of the foundation should be mopped with a hot bitumen until the whole surface is completely covered to a depth of at least $\frac{1}{16}$ inch. Drainage tile may be placed around the outside edge of the footing and a gravel fill, at least 12 inches deep, placed over the tile.

Laying the Tile. Tile lines, around the footings, should have a slope of about $\frac{1}{2}$ inch in every 12 feet. Proper slope can be judged by marking the sides of the footings. Then, gravel can be filled in accordingly. A long 2 by 4, laid along the tile lines, and a plumb level also help to establish ample and gradual slope.

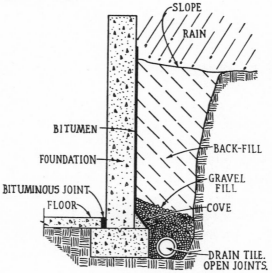

FIGURE 17. How to ensure dry basements.

At the low point, the bottom of the tile should be level with the bottom of the footing. The tile should be divided into two branches running from the low point in opposite directions around the footing to meet at the high point. Discharged water should be piped to a sump pump or other means of disposal.

Joints between the tile should be covered with pieces of roofing felt or tar paper to prevent soil from entering the tile, but still allow water to enter the joints.

All downspouts should be connected to drains or arranged to discharge water a considerable distance away from the foundations. Wherever possible, a sloping grade, as shown in Figure 17, should be provided so that rain and melt water from snow will drain away from the foundation.

INSULATING CRAWL SPACES*

Crawl spaces are apt to cause damp and cold floors, especially in climates where rain and cold weather are the rule. Such spaces can be insulated as follows:

Heavy roofing paper can be spread over the soil surface. The joints should be well lapped and sealed, using hot bitumen. Around the foundations, the paper should be sealed to the concrete, also by use of the bitumen. This method of insulation will eliminate much dampness and make floors more comfortable and safe from rot. Thick plastic sheets, with cemented joints, may be used instead of roofing paper.
or nailed to their underedges. Care should be taken to see that the moisture barrier faces the floor. This method of insulation will help to make the floors warmer.

FOUNDATION CONDENSATION*

It sometimes happens that the inside surfaces of basement walls (foundations) tend to "sweat," or become damp during the warm summer months. This happens when the warm and humid air in the basement comes in contact with the cool surfaces of the concrete foundations.

Warm air can and does entrain (carry) more water vapor than cool air. When such air comes in contact with the cooler surfaces of the foundation, it is cooled. At the lower temperature it cannot carry or hold as much water vapor, and the excess is deposited on the surface of the foundation. This causes the sweating, or dampness, so often noted.

As a means of reducing such condensation, basement windows can be kept open as a means of ventilation. Electric fans can be used to increase the ventilation.

In order to stop such condensation completely, the surfaces of the foundation must be insulated. This can be accomplished by attaching 1- by 2-inch strips (furring) to the foundation. The strips should be spaced on 16-inch centers. Any one of several plain or decorative kinds of 1-inch

* Also see Chapter 13, Insulation.

thick insulation can then be nailed to the strips. The insulation prevents warm and moisture-laden air from coming in contact with the cool surfaces of the foundation and thus stops the objectionable sweating.

FORM REMOVAL

During warm weather foundation forms can safely be removed 2 or 3 days after the concrete has been placed. However, at that time the concrete is still green and may chip easily. When there is no hurry, it is wise to leave the forms in place for several days. During cold weather the forms should not be removed for at least a week.

CHAPTER SEVEN

Concrete Floors

Well-built concrete floors provide the utmost satisfaction in terms of economy, durability, fire safety, wear resistance, and long-time service for a wide range of requirements in residential, farm, and other types of buildings.

Good planning, as previously mentioned in regard to other types of masonry construction, is essential and should be carefully done prior to the time the actual placing of concrete is started. A great deal of trouble and extra expense can be avoided if and when all floor projects are carefully studied and planned on paper. This is the only possible way of making sure that all such projects will properly serve their intended purposes.

Good construction procedures and workmanship are also highly important. These aspects of concrete-floor work may require a little more time and effort but pay big dividends in the long run. This is especially true where inexperienced masons are concerned.

The purpose of this chapter is to explain and illustrate several commonly encountered types of concrete-floor planning and construction.

COMMON CONSTRUCTION FAULTS

A brief discussion of the following important points will help inexperienced masons to prevent some of the more common construction faults which apply to all types of concrete floors:

105

Mixing Water. In Chapter 4, Concrete, special mention was made about the amount of water to use per bag of cement in any particular mix. It was pointed out that too much water reduces the strength and durability of concrete. This is especially true regarding floors because they are relatively thin layers of concrete which have great expanse. Some inexperienced masons have been led to believe that excessively wet mixtures are easier to place without unsatisfactory results. Quite the contrary! Sloppy mixes are most difficult to place without a great deal of risk so far as separating the cement paste from the aggregate is concerned. If the paste runs away, the strength of any floor is seriously impaired. Also, sloppy mixes are apt to cause shrinkage, crazing (cracking of the surface), and dusting (continued formation of cementlike dust long after the concrete has hardened). Furthermore, sloppy mixes require much more labor time and effort.

If the amounts of water specified in Table 1 of Chapter 4, Concrete, are used, none of the troubles just explained are likely to occur. It is highly important that masons understand this phase of concrete mixing because it means the difference between good work and unsatisfactory work. It means the difference between good craftsmanship and poor masonry work. It is also the difference between money well spent or wasted.

Wear Resistance. There are two general types of floors which most masons are apt to encounter: floors which are subjected to severe wear (as in factories) and floors (as in houses) which do not receive severe wear. Floors should be carefully planned and constructed to meet the needs they are to serve.

Factory Floors. When a floor is subjected to severe wear, more coarse aggregate should be used close to its surface. The coarse particles take the wear. For such floors, little or no surface troweling should be done so that particles of aggregate will remain close to the surface. In mixes for this type of floor, proper amounts of water are especially important as a means of preventing fine aggregate from accumulating at the surface.

House Floors. Where floor slabs on ground or basement floors are concerned, surface wear is not great, and a smooth surface is desirable. Thus, less coarse aggregate is needed near the surface.

Ordinary concrete slabs or basement floors can be made without the necessity for a topping composed of cement and sand. This saves both time and expense. The concrete can be placed all in one operation and worked to provide a smooth and satisfactory surface.

After the concrete is placed, it should be well spaded. This process tends to settle some of the coarse aggregate well below the surface. Also, an ordinary but heavy rake can be used to bring more fine materials closer to the surface. Immediately after the concrete is placed and leveled (by screeds), the rake, with its handle held in a vertical position, is used to tamp the concrete. The broad side of the rake should contact all surfaces of the concrete with repeated, rather heavy strokes. This procedure tends to push the coarse aggregate down and to allow more fine materials to accumulate at the surface. Such fine materials can be finished to a smooth surface which is completely satisfactory.

Especially made tools are available for tamping instead of the rake mentioned above. When masons expect to do considerable floor work, it will pay them to purchase such a tool.

Manipulation. Excessive troweling should be avoided on any type of concrete-floor surface. Otherwise, crazing and dusting are the results. The use of trowels, especially steel trowels, should be confined to just the amount absolutely necessary for proper smoothing. Inexperienced masons should strive to make every motion of a trowel count so that the operation can be reduced as much as possible. It should be kept in mind that the more troweling done, the more chance of highly objectionable crazing and dusting.

Troweling should be started only after all surface water has disappeared and the concrete looks dull. In other words, the trowel should be used after the concrete has started to stiffen. Any troweling done before this stage will simply constitute useless work and increase the chances of crazing and dusting.

Dust Coats. Concrete mixes will not be sloppy if they are proportioned as explained in Chapter 4, Concrete. Making a trial batch will avoid any chance of too mushy or too sloppy a mix. However, now and then a mix does become too wet through carelessness or a mistake in adding the water. When sloppy mixes do occur, they should never be dusted with cement or a mixture of cement and sand as a means of hastening the finishing work. For example, if the surface of a placed floor seems to remain wet and unworkable too long, it is poor policy to spread cement or cement and sand over its surface as a means of soaking up the water. This is one fault which has been made in all too many cases.

Such dusting is sure to cause crazing, and the surface will always give off a cementlike dust which makes any floor most undesirable. No matter how wet a mix might be, the finishing should not be started until the water disappears naturally.

Curing. If the surface of a concrete floor is allowed to dry out too soon, an inferior wearing surface usually results. Also, the relatively thin layer of concrete does not develop sufficient strength. Because of the small mass of concrete (thickness) and the large expanse of exposed surface, special care should be taken to keep the concrete wet until it is thoroughly cured. Improper curing causes rapid shrinkage at the surface, which results in crazing and dusting.

The foregoing procedures and explanations constitute excellent advice to all inexperienced masons. If the suggestions are followed, many of the otherwise bothersome troubles of masonry work can be avoided.

RESIDENTIAL FLOORS ON GROUND

For residences without basements or for unexcavated portions under them, concrete floors on ground provide an excellent type of construction. Studies have shown that, when properly planned and constructed, concrete floor slabs on ground give better results than other types of floors for basementless construction.

Basementless houses, regardless of the type of floor construction, should not be erected in low-lying areas that are damp or in danger of flooding from surface water. The surrounding ground area should slope away from the house with good drainage and should be at least 6 inches below the finished floor level. Thus, before any such house is planned, the site should be carefully investigated because -dampness can absolutely ruin the house as a place to live. It is wise to investigate the history of the site so far as surface water is concerned.

Subgrade. It is important that the subgrade be well and uniformly compacted to prevent any unequal settlement of floor slabs. In other words, as repeatedly stressed, the concrete must be amply supported.

All organic matter such as sod, roots, and weeds should first be removed and the ground leveled off. Any holes or irregularities in the subgrade and any trenches for utilities should be filled in layers not more than 6 inches deep and thoroughly tamped. If holes or trenches are filled all in one operation, instead of in shallow layers, settlement is sure to occur. Material used for fill should be of uniform character and should not contain large lumps, stones, or frozen chunks.

The entire subgrade should be rough-graded to an elevation slightly above the finished grade and then carefully compacted by tamping or rolling. This procedure is important! Unless it is done properly, floor cracking can occur. The finished subgrade should be carefully checked to make sure it is at the elevation desired.

For the best compaction results there is a proper moisture content for each type of soil. A rough idea of the proper moisture content of ordinary soils, except extremely sandy ones, can be obtained by squeezing some in the hand. With proper moisture content the soil can be molded but will not be muddy. If the soil is too dry, it should be sprinkled with water and mixed to the proper condition before compacting is started. If the soil is too wet, it must be allowed to dry before compacting is started. In sandy soil water can be added freely because any excess will not be harmful.

Fill. A coarse fill, generally composed of gravel or crushed stone, as shown in Figure 1, should be placed over the finished subgrade. The fill should be thoroughly tamped and compacted. Fills are intended to serve both as an insulation material and as a protection against moisture from the ground. The fill material should be uniform-sized particles, without dust, so as to ensure a maximum volume of air space (voids) in the fill. If necessary, the material should be screened to remove all fine material. The voids will add to the insulating ability of the fill and resist the passage of subsoil moisture.

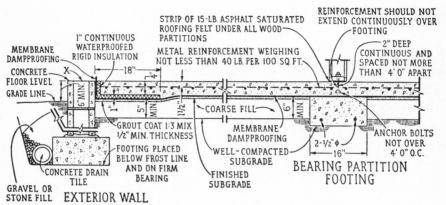

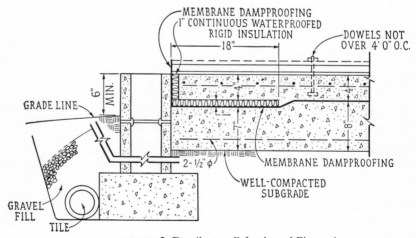

FIGURE 1. Typical construction of concrete floor on ground.

FIGURE 2. Detail at wall footing of Figure 1.

The exception to the above fill requirements applies to warm dry climates. For example, Western practice allows the use of dry sand as fill material. In such climates the sand serves the purpose satisfactorily.

A line of draintile, as explained in Chapter 6, Concrete Foundations, may be placed, as shown in Figure 2, around the outside edge of the foundation or footings and connected to a sump pump or otherwise drained to minimize the possibility of soil moisture entering the fill. Any moisture in the fill will reduce its insulating ability. Tile drains are not necessary where houses are to be located on high ground or where the soil is naturally well-drained or dry.

Provision for Utilities. Where required, provisions can be made for plumbing lines under the fill. If water-supply lines are placed under the fill, they should be installed in trenches of the same depth as those outside the building. Connections to such lines can be brought to a point above the concrete floor level prior to the time concrete is placed.

Dampproofing. After the fill has been compacted, a stiff grout (see Figure 1), consisting of one part cement and three parts sand, should be placed over it to provide a smooth surface for installing membrane damp-proofing (see Figure 1). The grout should be at least ½ inch thick and should be broomed (swept) into place. The grouted surface when hardened and dried and also the top of any partition footing (see Figures 1 and 2) should be mopped with hot bitumen. As the mopping proceeds and before the bitumen has time to cool and harden, a layer of 15-pound asphalt-saturated roofing felt should be placed over the bitumen with the edges of the felt well lapped. Two layers of the felt are recommended with hot bitumen mopped between layers and on top of the second layer. This membrane dampproofing should be continuous over the entire floor area and carried up the inside of the foundation, as shown in Figure 1, to a point 1 inch or more above the finished-floor level.

Insulation. As shown in Figures 1 and 2, a 1-inch thick continuous layer of waterproofed rigid insulation should be provided between the foundation and the edge of the floor slab. This insulation is highly important in terms of comfortable floor temperatures.

Reinforcement. EASTERN PRACTICE. Metal reinforcement, in the form of woven wire, as shown in Figure 1, weighing not less than 40 pounds per 100 square feet, should be placed in the concrete slab 1½ inches below the surface.

WESTERN PRACTICE. In regions where earthquakes are apt to occur, most building codes require that ½-inch steel bars be placed in the concrete and spaced about 4′ 0″ on center. The rods are placed parallel and should extend the full length and width of the slab. Readers are urged to check their local codes because reinforcement requirements vary widely.

Bearing-partition Support. Partitions which help to support floor or roof loads are known as *bearing* partitions. Such partitions have to be supported in order to prevent settlement and cracks.

Figure 1 shows how bearing partitions must be supported by footings which should be at least 16 inches wide and 8 inches deep. Such minimum dimensions should be confined to ordinary small houses and other buildings. In cases where loading is apt to be more than common to such buildings, the design method explained in Chapter 5, Concrete Footings, should be used.

Note that the sills for wood partitions should be held in their proper positions by means of anchors which are cut (embedded) into the concrete floors.

Figure 1 indicates that at least two steel bars are recommended near the bottoms of bearing-partition footings. In some earthquake regions building codes call for more than two and sometimes for larger bars.

RESIDENCE FLOOR SLABS FOR RADIANT HEATING

When concrete floor slabs on ground are to contain radiant-heating pipes, careful attention should be given to the matter of additional insulation.

The *A* part of Figure 3 shows a typical radiant-heating installation. In principle, the hot water circulated through the pipes heats the concrete slab, and it in turn radiates heat to the interior. It can be readily understood that loss of heat to the soil, around the exterior of the foundations, is most undesirable. The curving dashed lines show the approximate path taken by escaping heat.

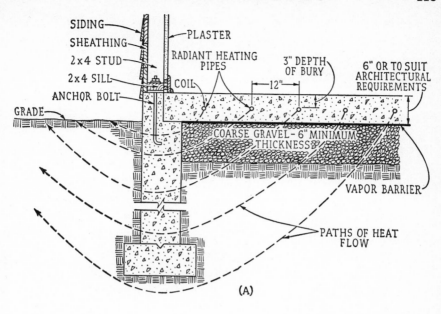

SIDING
SHEATHING
2x4 STUD
2x4 SILL
ANCHOR BOLT
GRADE
PLASTER
RADIANT HEATING PIPES
COIL
3" DEPTH OF BURY
12"
6" OR TO SUIT ARCHITECTURAL REQUIREMENTS
COARSE GRAVEL - 6" MINIMUM THICKNESS
VAPOR BARRIER
PATHS OF HEAT FLOW

(A)

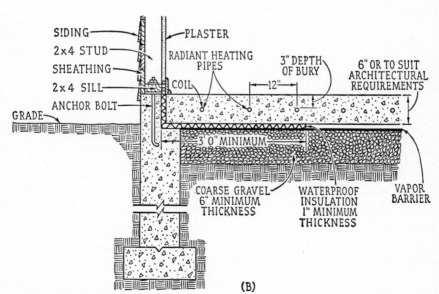

SIDING
2x4 STUD
SHEATHING
2x4 SILL
ANCHOR BOLT
GRADE
PLASTER
RADIANT HEATING PIPES
COIL
3" DEPTH OF BURY
12"
6" OR TO SUIT ARCHITECTURAL REQUIREMENTS
3' 0" MINIMUM
COARSE GRAVEL 6" MINIMUM THICKNESS
WATERPROOF INSULATION 1" MINIMUM THICKNESS
VAPOR BARRIER

(B)

FIGURE 3. Improper (*A*) and proper (*B*) methods of insulating concrete slabs when radiant heating is used.

The *B* part of Figure 3 shows a recommended use of waterpoooof insulation which resists heat loss. This type installation of insulation has been found to be effective and ensures that most of the heat will be radiated into the interior of the building.

PLACING A CONCRETE SLAB ON GROUND

As an example of the procedure which inexperienced masons may follow in placing a concrete slab on ground, without the use of surveying and other special instruments, the following situation is assumed:

Figure 4 shows the concrete-block foundation, *ABCDEF,* for a small house for which the exterior walls above grade are also to be constructed of concrete block. The house is to have one bearing partition which will extend from the corner *C* to the side *AF*. The foundations, as shown in Figure 1, are already in place.

Footing Location and Level. The plan in Figure 4 shows that the center line of the bearing partition is to be 7' 8" from the exterior face of the *AB* side of the foundation. The footing under that partition must be 16 inches wide and 8 inches deep.

The center line of the partition can be indicated by stakes and string between points *s* and *t*.

Once the center line is established, the forms for the footing can be assembled and placed, subject to determining their exact level. They can be made of 2 by 8s as explained in Chapter 5, Concrete Footings.

The top surface of the footing, as shown in Figures 1 and 2, must be exactly 4 inches below the surface of the floor and the same distance below the top of the foundation. In order to establish the top edges of the forms accordingly, a string can be tightly stretched from the positions of *m* and *n,* as shown in Figure 4. The string can be held by stakes so that it just touches the tops of the foundation at *m* and *n*. The top edge of the forms should be adjusted so that it is exactly 4 inches below the string. The string can be moved to several locations along the forms, so that their proper level can be established from end to end.

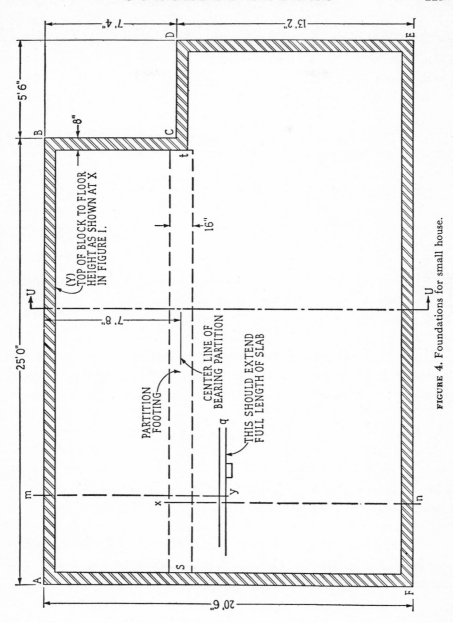

FIGURE 4. Foundations for small house.

The form level can be checked using the following procedure:

The *W* part of Figure 5 shows a section taken at *U-U* in Figure 4. The two sides of the foundations and the footing forms are shown. On the foundation, make a pencil or chalk mark at *x,* which is exactly 4 inches below the top. Then place one end of a straight 2 by 4 across the forms, as shown, and hold the other end so that its bottom edge is even with the mark at *x.* Place a level on the top edge of the 2 by 4. If the level bubble is in proper position, the tops of the forms are at the correct elevation. If not, adjust the forms accordingly. This procedure should be repeated at intervals along the forms. If done carefully, it is a suitable procedure.

It is desirable and helpful to place the footing reinforcement and concrete at this stage so that finished footing surface can be used as explained in the following:

Finished Grade. The finished grade should be prepared as explained earlier in this chapter. Its level can be checked according to the procedure indicated in the *X* part of Figure 5.

Make a mark on the foundation, as shown at *x,* exactly 10 inches below its top. The finished subgrade must be that distance below the top of the foundation because the floor and fill must be 4 and 6 inches deep, respectively. Next make a mark on the footing, as shown at *x,* which is exactly 4 inches below its top surface. Use a piece of 2 by 4 and a level as shown. This same procedure can be repeated at several points along the subgrade, on both sides of the partition footing, until all necessary adjustments to the subgrade have been made. Also, a piece of 2 by 4 can be placed at random on the subgrade and a level used to make sure that all areas of it are at the same level.

Fill. The fill should be placed to approximate thickness as explained earlier in this chapter. Its level can be checked according to the procedure indicated in the *Y* part of Figure 5.

Make a mark on the foundations, as shown at *x,* which is 4½ inches below its top. Also make a mark on the footing, as shown at *x,* which is ½ inch below its top surface. Use a piece of 2 by 4 and a level, as shown, to check the level of the fill. Repeat the operation at intervals on both sides of the footing. Also, a piece of 2 by 4 can be placed at random on

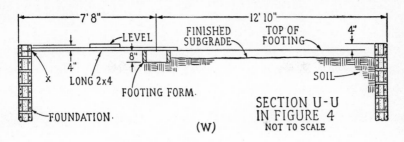

SECTION U-U
IN FIGURE 4
NOT TO SCALE

(W)

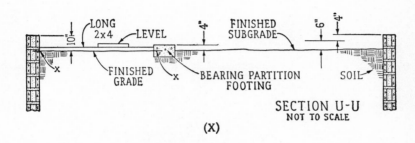

SECTION U-U
NOT TO SCALE

(X)

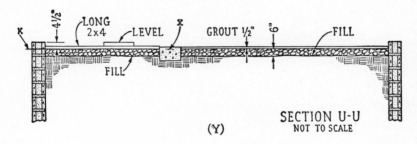

SECTION U-U
NOT TO SCALE

(Y)

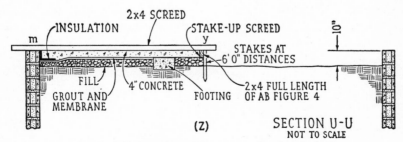

SECTION U-U
NOT TO SCALE

(Z)

FIGURE 5. Leveling without surveying instruments.

the fill and a level used to make sure that all areas of it are at the same level.

Dampproofing. Mix and place ½ inch of the grout as explained earlier in this chapter. Figure 6 shows a mason applying a second layer of felt.

FIGURE 6. Applying second layer of roofing felt. (Courtesy of Portland Cement Association.)

Insulation. The insulation should next be placed as explained earlier in this chapter. It will adhere to the bitumen.

Placing Concrete. In order to place and level the concrete to advantage, a staked-up screed, such as shown in the Z part of Figure 5, should be erected. The correct level for the screed (also indicated as q in Figure 4) can be established by stretching another screed, *my,* so that it rests on the top of the foundation and on the staked-up screed at y. Put a level along *my* and adjust the height of the staked-up screed as may be necessary. This procedure should be repeated at intervals along the full length of the staked-up screed, from side *AF* to side *DE* of the foundation. In Figure 4, the screeds between the staked-up screed and the foundations are indicated by lines *my* and *xn*.

Place concrete on only one side of the staked-up screed at a time and to a first depth of $3\frac{1}{2}$ inches. Then place the reinforcement and the balance of the concrete.

Once the concrete is all placed to approximate depth, use one of the screeds, such as *my* or *xn*, to level it. The screed should be handled by two men so that it can be zigzagged back and forth. The ends of the screed must touch the foundation and the staked-up screed. Then, as the zigzagging is continued from one end of the floor to the other, the concrete will be spread at the proper thickness under the screeds *my* or *xn*. The *Z* part of Figure 5 indicates this.

Place concrete on the other side of the staked-up screed and spread it in the same manner. Then, remove the staked-up screed and fill the furrow that it left with concrete.

Finally, use a 2 by 4 screed, which is somewhat longer than the distance between foundation sides *AB* and *FE*, to zigzag over the whole floor again.

If the foundations at points *s* and *t* as shown in Figure 4, are marked prior to concrete placing, a string can be stretched between *s* and *t* which will indicate the center line of the footing. The anchor bolts, as shown in Figure 1, can then be shoved down the proper distance in the concrete. Care should be taken to make the anchors perfectly vertical.

A rake, as previously explained in this chapter, can be used to make the surface of the concrete suitable for wood and steel troweling.

Before using the wood trowel (often called a *float*), all surface water should be allowed to disappear naturally, and the concrete should be starting to stiffen. Use the wood trowel to finish the surface roughly and to fill all depressions, etc. Finally, use the steel trowel just enough to give the floor a smooth texture. Troweling procedure can only be learned by actual practice. Inexperienced masons should not feel badly if their first efforts are not works of art.

FINISHES FOR ON-GROUND SLABS

Practically any type of finish or finish materials can be applied to such a slab. Common practice is to use hardwoods, over-all carpeting and rugs in living rooms; linoleum, asphalt tile, ceramic tile, and similar coverings in kitchens, bathrooms, and halls.

Wood Covering. The usual method of putting down wood floors on concrete slabs is to nail them to wooden sleepers which are attached to or embedded in the concrete. The *X* part of Figure 7 demonstrates the use of 2 by 2 wood sleepers which can be attached to the concrete by means of clips or by using special nails which will penetrate concrete. The *Y* part of the illustration shows the use of embedded sleepers. In either case subflooring and finished flooring or just finished flooring may be applied.

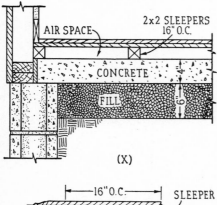

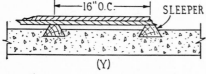

FIGURE 7. Details of wood sleepers for use when wood finish is applied over concrete slabs.

Carpeting and Rugs. Carpets and rugs are customarily laid over pads which add to the life of the covering and give a sensation of walking on a deep-piled rug. Over-all carpeting is usually stretched and tacked to wooden strips around the edge of the floor. These can be embedded in the concrete or nailed to wood plugs set in the concrete.

Linoleum-type Coverings. Concrete slabs make an excellent foundation for linoleum and various kinds of soft tile. These materials are cemented directly to the concrete surface using an adhesive which manufacturers supply.

Ceramic Tile. A concrete slab is required for the proper setting of such tile. The concrete should be given a rough finish or a coarse, broomed texture to provide a good bond for the mortar used in setting the tile. Just before setting, a slush coat of *neat* (pure cement without sand) cement grout should be broomed into the surface of the concrete floor. Before the grout hardens, it should be covered with a mortar-setting bed about ¾ inch thick. The mortar should consist of one part cement and three or four parts sand. Only as much mortar is spread at one time as can be covered with the flooring material before the mortar begins to harden. The floor tile should be placed upon and tapped into the setting mortar until true and even. The joints between tile should be filled with a mortar composed of one part cement, not more than one part sand, and the required mixing water. The mortar must be forced into the joints until they are well filled. All surplus mortar should be quickly removed.

Color Finishes. Concrete floor slabs can be colored by incorporating mineral pigment into the concrete. Or when especially beautiful surfaces are required, a colored topping composed of one part cement and three parts of sand can be used. See Chapter 3, Mortar, for information about colors.

BASEMENT FLOORS

As an example of the procedures which inexperienced masons may follow in placing a basement concrete floor without the use of special equipment, the following situation is assumed:

Figure 8 shows the basement plans for the house indicated in Plates I through VI. The chimney and floor drain are in the proper positions. The footings for the foundations and for the chimney are as illustrated in Plates II and VI.

With the floor drain and all the footings in place, the soil can be leveled by the use of long 2 by 4s. The soil level should be 2 inches lower than the top surface of all footings and 4 inches lower than the floor drain. The area of the floor to the right of the chimney should have a slight slope toward the floor drain. It can be assumed that the plumbers have already set the drain properly.

In order to place and spread the concrete easily, several screeds, as indicated by the dashed lines in Figure 8, may be used. The surface of

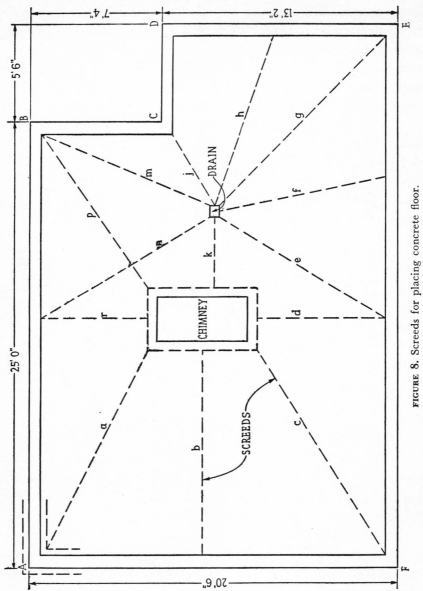

FIGURE 8. Screeds for placing concrete floor.

the concrete floor must be 2 inches above the footings and flush with the top of the drain. Therefore, the screeds which contact the footings should have their ends cut, as shown in Figure 9.

The various screeds may be set as indicated by the dashed lines in Figure 8. Those screeds which contact the drain should have normal ends set with their top edges flush with the tops of the drain.

Place the concrete, for example, between screeds *a* and *b* first. Then use another piece of 2 by 4 from screed *a* to screed *b* so that the concrete can be leveled as explained previously for the concrete slab on ground.

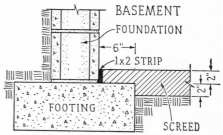

Next, place concrete between screeds *a* and *r* and level with another piece of 2 by 4 screed. When these two sections have been placed and leveled, screed *a* can be removed and the furrow it left filled with concrete.

FIGURE 9. Details of end of screed.

The same procedure can be followed for all other sections of the floor. The finish operations are the same as explained previously for the concrete slab on ground.

FINISHES FOR BASEMENT FLOORS

Concrete basement floors can be finished in various ways. The concrete can simply be troweled smooth and left in its natural color, or it may be given a linseed-oil treatment and polished with wax.

If a special colored topping is desired, such as for a recreation room, the regular concrete can be placed only to a depth of 3 inches and a topping mixture, composed of one part cement and three parts sand, placed to a depth of 1 inch. (The placing of topping is explained in Chapter 15, Miscellaneous Concrete Projects.) The coloring is accomplished as explained in Chapter 3, Mortar, and Chapter 4, Concrete.

Another method of coloring is to use a dusted-on color mixture. After screeding the floor, a mixture of approximately one part cement, one to one and one-half parts sand, and the required amount of color pigment (see Table 1 of Chapter 3, Mortar) can be dusted uniformly over the fresh concrete. This dry mixture should be worked into the concrete, using a wood trowel, until the surface becomes wet, and resumed after the water disappears naturally. Finally, a steel trowel can be used to do the final smoothing. NOTE: This coloring procedure is apt to cause some crazing or dusting.

Wood, tile, etc., can also be used as previously explained.

RESURFACING OLD FLOORS

If existing old floors become damaged in one way or another or are not smooth or all at the same level, they can be resurfaced with a bit of tedious work and care.

The old concrete surface must be broken out to a depth of at least 1 inch. This can be accomplished by the use of a power hammer. It can also be done using a cold chisel and a heavy hammer. An entirely new and rough surface should be exposed.

The new surface must be thoroughly cleaned using a stiff wire brush and water so that it is entirely free from soil, grease, and all loose material.

A new topping of one part cement and three parts sand, just plastic enough to be spread with ease, can be placed to a depth of 1 inch or more. The new surface should be moist before the topping is applied. The finishing can be done as previously explained.

Old concrete floors can also be patched in spots where they have been damaged. However, patched spots never have a good appearance, and there is difficulty in determining the exact thickness to make the patch. The mortar material shrinks as it hardens, and the patched spot is apt to be lower than other parts of the floor, as well as being a different color.

PAVED AREAS

As an example of the procedures which inexperienced masons may use in placing concrete for paved areas, without the use of special tools, the following suggestions are presented:

Subgrade. The subgrade must be level. The leveling process can be accomplished by establishing some desired elevation and using methods similar to those explained for Figures 4 and 5. The soil should be firm and well tamped wherever it has been disturbed. If the soil is apt to be wet or poorly drained, a 6-inch fill of gravel or crushed stone is recommended.

Planning. Most paved areas are made 4 inches thick for normal use. However, where heavy trucks or machinery might be drawn over the areas, the concrete should be at least 6 inches thick and should include some reinforcement, as local building codes may specify. A slope should be provided so that water will drain in a desired direction.

Edges of paved areas may be thickened underneath or raised, as shown in the *W* and *X* parts of Figure 10, to provide greater strength. Thickening the edges helps to prevent undermining, and raised edges help to divert water to desired locations such as drains.

Construction Joints. If paved areas are to be more than 10 feet wide, the concrete should be placed in about 10-foot strips, leaving every other strip unpaved until the first strips have hardened. The joints between strips, as shown in the *Y* part of Figure 10, should be given careful consideration. The tongued-and-grooved joint permits some movement but prevents cracking or changes in the general level of the paving surface.

Forms. Each strip of pavement should have 2 by 4 or 2 by 6 forms which can be set in place following the general procedure outlined in Chapter 5, Concrete Footings. All form pieces must be level. This can be checked by placing a level on them and on a 2 by 4 stretched from one form to the other. Figure 11 shows the forms for one strip of pavement.

Placing Concrete. The concrete should be placed and spread to approximate level by the use of a shovel. Then a screed, such as is also shown in Figure 11, is used to create the final level.

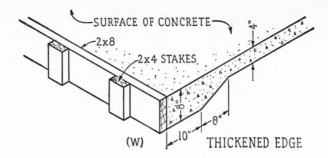

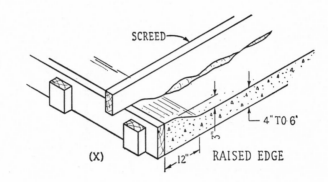

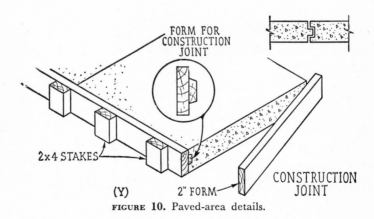

FIGURE 10. Paved-area details.

FIGURE 11. A screed being used to level one strip of pavement. (Courtesy of Portland Cement Association.)

PORCHES

As an example of the procedures which inexperienced masons may follow in placing concrete for porches and accompanying trench walls, without the use of surveying or other special tools, the following situation is assumed:

Plates I, II, and IV indicate that there is to be a concrete-slab porch just outside of the service room. The plates also indicate that the north and east sides of the slab must be supported by trench walls which are 8 inches thick and 3′ 4″ deep without footings.

In order better to visualize what concrete work is necessary and how the forms must be set, Figures 12 and 13 should be studied.

Plate IV indicates that the house floor level is to be 1′ 3″ above the grade line and that the porch must also serve the purpose of two risers in that vertical distance. The risers will be located at the x and y positions in Figure 12. In order to make each of the risers of comfortable height, the top surface of the porch floor can be planned, as shown in both Figures 12 and 13, to be 7½ inches below the surface of the house floor. Then, the risers at x and y will both be equal to 7½ inches because the distance between the house floor level and the grade line is 15 inches. A careful study of Figures 12 and 13, together with Plates I, II, IV, and VI, will make the situation clear.

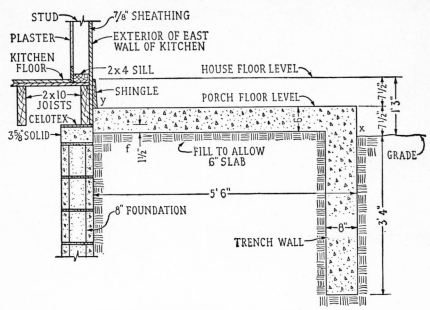

FIGURE 12. West and east section of kitchen porch.

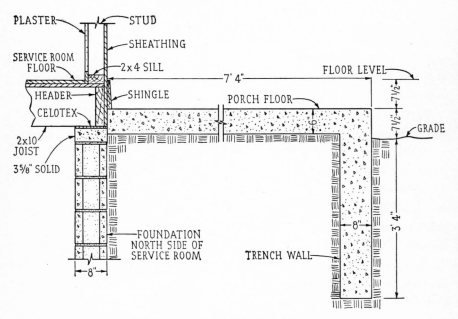

FIGURE 13. North and south section of kitchen porch.

Forms. Figure 14 shows one method by which the necessary forms can be provided. Assuming that the soil will stand erect to form the sides of the trench wall, as shown at *a* and *b,* only the wood forms shown are necessary.

Forms *c* and *d* can be made, using pieces of 2 by 8 material. These two pieces must be placed so that their lower edges, as shown at *e,* are about

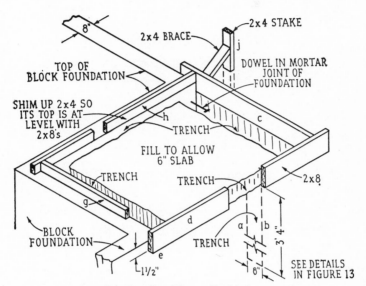

FIGURE 14. Details of forms for kitchen porch.

1½ inches below the top of the foundation. This is also shown at *f* in Figure 12.

The *g* and *h* form pieces can be made of 2 by 4s if they are shimmed up to the level of the *c* and *d* pieces. Pieces of wood shingles or slivers of wood can be used as shims.

A level should be used to see that all forms are level and all are at the same level. This can be accomplished by placing a level along the top edge of each piece and by then stretching a 2 by 4 across the forms and putting the level on the 2 by 4.

Because the pieces *c* and *d* are both right at the edge of the trench, stakes cannot be driven without disturbing the sides of the trench. Thus, stakes and braces that are somewhat removed, such as those shown at *j,* must be used.

Placing Concrete. Place the concrete in the trenches as explained in Chapter 6, Concrete Foundations. Then fill the space between the forms. Spade the concrete well, especially next to the forms and in the trenches. A screed should be stretched across the forms for leveling purposes. A rake may be used, as previously explained, to settle coarse aggregate somewhat below the surface of the concrete. Wood and metal trowels can be used in the usual manner. An edger should be used to round the corners next to forms *c* and *d*.

When the concrete has stiffened to the point where it feels hard to the light touch, form pieces *c* and *d* can be carefully removed so that the vertical surfaces back of them can also be troweled smooth.

PATIO FLOORS

In many instances, concrete patio floors are desirable. They can be made any shape and, by the use of 2 by 4s, in many pleasing patterns.

Figure 15 shows a typical patio floor which is 12 feet square and in which 2 by 4s are used to create the pattern or grid.

Construction of Grid. Well-dried 2 by 4s should be used so as to prevent warping after the concrete has been placed between them. The various pieces can be cut before assembly is started. For example, piece *AB* should be exactly 12′ 0″ long. Piece *BC* must be 12′ 0″ minus the thicknesses of two 2 by 4s. Piece *SN* is 3′ 0″ minus the width of one 2 by 4.

Assemble the grids using 3-inch nails where butt joints are concerned and somewhat smaller nails where toenailing is necessary.

In order to prevent cracks between the 2 by 4s and the concrete, 4-inch spikes should be driven, at about 6-inch intervals, as shown in the *Y* and *Z* parts of Figure 16.

Give the grid of 2 by 4s one or two coats of shellac prior to the time it is placed in final position.

The *X* part of Figure 16 shows the necessary excavation. A fill should be provided where the soil is apt to be wet and where winter freezing occurs. Test the level of the bottom of the excavation by placing a 10- or 11-foot 2 by 4 at random and using a level on the top edge of it.

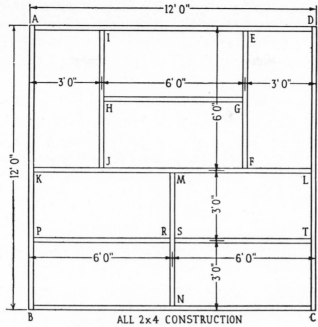

FIGURE 15. Typical design for patio floor.

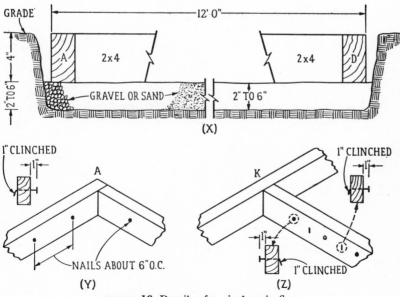

FIGURE 16. Details of typical patio floor.

If fill is used, it should be well tamped and tested for level. If sand is used as the fill, it can be wetted and thus leveled to better advantage.

When the grid is placed in the excavation, care should be taken to see that the top edges of the 2 by 4s are at the exact grade desired and that the whole grid is level.

Place the concrete (plain or colored) in one section of the grid at a time. Use a screed to level and a trowel to smooth the concrete. Remove concrete from the top edges of the 2 by 4s as quickly as possible. The exposed edges of the 2 by 4s can be painted or left in their natural color.

CHAPTER EIGHT

Concrete Block

When portland cement, water, and suitable aggregates, such as sand, gravel, crushed stone, cinders, burned shale, or slag, are mixed and formed into individual pieces to be used in laying up walls and other structural details, the pieces thus formed are known as *unit masonry,* or *units.* Such units are also technically known as *concrete masonry.* However, most masons use the terms *concrete block* or *block* when they refer to such material. In this book, the terms *concrete block, block,* and *units* are employed interchangeably and to mean the same material.

Concrete-block construction is of great interest to masons and to others who are concerned with the planning and erection of economical, sound structures of all kinds. This type of masonry permits easy planning and quick erection and can be done by masons having little previous experience.

The purpose of this chapter is to explain the fundamentals of concrete-block construction and to prepare readers for subsequent chapters in which actual construction procedures are explained.

CHARACTERISTICS OF CONCRETE BLOCK

Concrete block of good quality assures sound and long-lasting structural work of all kinds. Blocks with sharp, straight edges and corners are easily laid and are neat in appearance. Strength, dryness, and other characteristics of the units should meet the specifications of the American Society for Testing and Materials. Readers can be assured of first-class

133

block by dealing with a manufacturer having a reputation for making quality products.

Solid and Hollow Block. A solid concrete block is defined as a unit in which the core area (openings or cells) is not more than 25 per cent of the gross (over-all) cross-sectional area. A hollow concrete block is a unit having a core area greater than 25 per cent of its gross cross-sectional area. Generally, the core area of hollow block will be from 40 to 50 per cent of the gross area. This is the greatest core area acceptable in ordinary concrete-block work. In this book, it is assumed that hollow units are used.

Heavyweight and Lightweight Block. Units are made with either heavyweight or lightweight aggregate and are known accordingly. A hollow concrete block of the 8 by 8 by 16-inch size will ordinarily weigh from approximately 40 to 50 pounds when made with heavyweight aggregate and from 25 to 35 pounds when made with lightweight aggregate. Heavy units are made with such aggregate as sand, gravel, crushed stone, and slag. Lightweight units are made with coal cinders, shale, and natural lightweight materials, such as volcanic cinders, pumice, and scoria. Both types of units may be used for all kinds of masonry construction. The lightweight units generally provide the best insulation against the passage of heat and cold through walls and other structural details in contact with rooms of houses and other buildings.

Strength. A typical building-code requirement (which is subject to modification by local codes in cities and towns) specifies that the maximum allowable compressive stresses in walls and foundations constructed of hollow units are 70 pounds per square inch when the walls are laid up with the mortars shown in Table 2 of Chapter 3 for ordinary service and 85 pounds per square inch for the mortars shown for walls subject to extremely heavy loads. This fact is most important, and readers are urged to keep it in mind at all times.

The allowable wall stresses provide an ample factor of safety when using units which have, for example, a compressive strength of 700 pounds per square inch. Tests carried out by the Portland Cement Association have established the relationship between the compressive strength of hollow concrete block and the walls or foundations in which they are laid up. These tests disclosed the fact that the compressive strength of loaded walls or foundations in pounds per square inch was approximately 42 per cent of the compressive strength of the unit when laid up with *face,* or *shell* (these terms are explained in succeeding pages) mortar *bedding* (the mortar joints blocks are laid in) and 53 per cent when laid up with *full mortar bedding.* (Full mortar bedding is illustrated in Chapter 10, Concrete-block Walls, Foundations, and Pilasters.)

Thus, a wall or foundation constructed of hollow concrete block having a minimum compressive strength of 700 pounds per square inch, when laid up with face-shell mortar bedding, will have a compressive strength of 0.42×700, or 294 pounds per square inch. Using an allowable compressive stress of 70 pounds per square inch, a wall or foundation having a strength of 294 pounds per square inch will have a safety factor of 4.2. This exceeds the safety factor of 4 which is generally considered ample for such masonry construction.

Sound-absorbing Value of Concrete Block. Tests have indicated that, if concrete blocks have an open texture on their exterior faces, they will absorb sound readily. Sound waves, upon striking a surface, are partly reflected, partly absorbed, and partly transmitted in varying amounts, depending upon the character of the surface. A smooth, dense surface, such as plaster or glass, will absorb only about 3 per cent of the sound that strikes it. Ordinary concrete block, unless especially surfaced for extreme smoothness, will absorb between 18 and 68 per cent of the sound striking it. The lightweight block absorbs much more sound than the heavyweight variety.

Dry Units. In common with many other building materials, concrete blocks shrink slightly, with a loss of moisture down to an air-dry condition. If moist units are placed in a wall or foundation, this natural shrinkage is apt to cause cracks. When concrete-block walls or foundations will be exposed to low relative humidities, such as are found in the interiors of heated buildings or in areas of the country which have dry climates, it is desirable that the block, at the time of laying, be dried to approximately the average air-dry condition to which the finished walls or foundations will be exposed.

When delivered, concrete blocks should be dry enough to comply with the conditions of the job. They should be maintained in such a dry condition by stockpiling them on planks, free from contact with the ground. They should be covered with roofing paper or tarpaulin for protection against wetting. Moreover, at the stoppage of work at any time, the tops of walls and foundations should be covered to prevent rain or snow from wetting them or entering the cores of the block. While such precautions may seem to cause tedious work, they often make the difference between a good and a poor job.

Concrete blocks should never be wetted immediately before or during the time they are being laid up in a wall or foundation. When the blocks are dry, the mortar adheres to better advantage, and much better joints are possible.

SIZES AND SHAPES OF CONCRETE BLOCK

Concrete-block sizes are usually referred to by their nominal dimensions. Thus, a unit measuring 7⅝ inches wide, 7⅝ inches high, and 15⅝ inches long is known as an 8- by 8- by 16-inch unit. In other words, the 8- by 8- by 16-inch size is the *nominal* size and not the actual size. When laid in a wall or foundation or other structural detail, with ⅜-inch mortar joints, the block will occupy a space exactly 16 inches long and 8 inches high.

Typical Sizes and Shapes. Figure 1 shows typical sizes and shapes of concrete block. It should be understood that both heavyweight and lightweight block can be obtained in these shapes. It should also be understood that other sizes and shapes are available, as will be indicated in subsequent chapters. Practically every manufacturer of block has his own shapes of block. However, the over-all dimensions are exactly the same. This will become apparent as other chapters in this book are studied.

Upon request, the Portland Cement Association will supply names and addresses of manufacturers in various localities. The manufacturers will

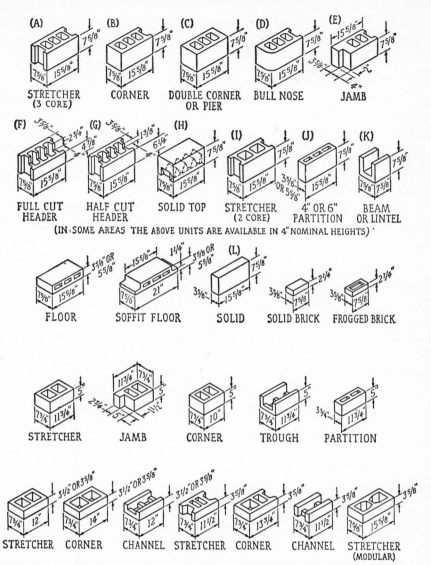

FIGURE 1. Typical shapes and sizes of concrete block. Dimensions shown are actual unit sizes. A $7\frac{5}{8}'' \times 7\frac{5}{8}'' \times 15\frac{5}{8}''$ unit is commonly known as an $8'' \times 8'' \times 16''$ concrete block. Half-length units are usually available for most of the units shown below. See concrete products manufacturer for shapes and sizes of units locally available.

in turn supply catalogues which show a great variety of two- and three-core blocks which can be used for many different structural purposes. In this and in subsequent chapters, several different sizes and shapes of block will be illustrated with typical structural details.

Readers are urged to obtain information about sizes and shapes of block locally available *before* they start to plan walls, foundations, or other structural details. As shown in Figure 1, units are usually described by their actual dimensions.

Most sizes and shapes of block can be secured in various colors, as well as in the standard gray of normal concrete.

Special blocks can be secured which have, for example, faces composed of marble chips. Such units are also manufactured in many colors.

USES OF CONCRETE BLOCK

Concrete-block construction can be successfully and economically used for many kinds of medium-sized buildings, as well as for chimneys, fire-places, piers, columns, pilasters, retaining walls, garden walls, and a great many other uses. Some of the typical uses are briefly explained in this chapter and further explained in subsequent chapters.

Concrete-block Homes. A large variety of attractive wall finishes can be applied to concrete-block homes. Interesting interior patterns can be worked out by tooling mortar points. Or attractive designs can be created with blocks of different sizes and shapes. For homes, block construction is reasonable in cost, and the building can be laid up very quickly. Low upkeep expense is another desirable characteristic of such construction. Walls and foundations withstand years of weathering without losing their pleasing appearances. In fact, age often adds beauty to such construction.

Fire safety is another desirable feature. A concrete-block wall or foundation will resist fire for a long period and will prevent its spread. Because of the insulating value of block, homes thus built are warm in winter and cool in summer. They can be erected in any climate with equal advantage, and they will assure pleasant and comfortable living.

Concrete-block Farm Buildings. The builders of farm buildings look for four essential qualities when selecting the materials with which to build: economy, fire safety, durability, and attractiveness. Concrete blocks meet these requirements and are readily adapted to farm-building needs.

Retaining Walls. Banks of earth, found on sites for homes or barns, can be retained in any desired position by the use of easy-to-lay walls which, when properly reinforced, prevent sliding and bowing out. Such walls can be made so that they are attractive in appearance and a pleasing addition to any yard or other area.

Other Walls. Ordinary fences, made of wood or wire, either require a great deal of maintenance or fail to provide privacy. A block wall, of any desired color and topped off with interesting trim, constitutes a permanent installation which requires little, if any, maintenance, provides complete privacy, and becomes one of the decorative points of a garden.

CONCRETE-BLOCK FINISHES

A wide variety of finishes are available for concrete-block construction. The finish to use in any given case should be governed by the type of structure in which the material will be used, the climatic conditions to which a structure will be exposed, the architectural effects desired, and personal taste. Several popular finishes are described in the following:

Architects have worked out many interesting variations in treatment, several of which are shown in Figure 2. The units may be laid in regular courses (rows) of the same height or in courses of two or more different heights, or several sizes of units may be laid up in a prearranged pattern. NOTE: Before deciding upon any particular pattern, especially those of the stacked variety, it is best to check with local building codes to make sure that the desired pattern is acceptable.

In some wall treatments, all the joints are accentuated by means of deep *tooling* (this term is explained later in this chapter). In others,

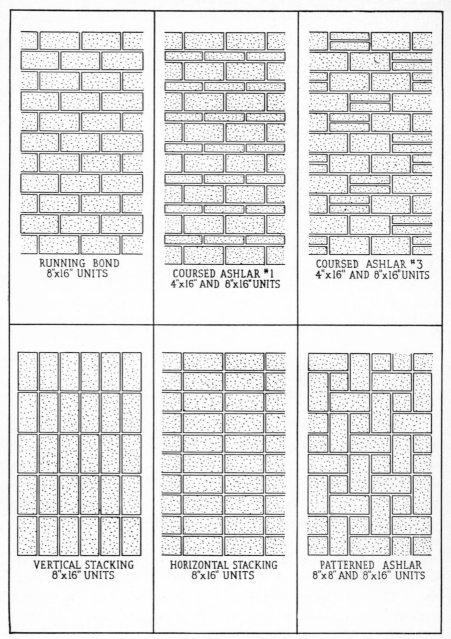

FIGURE 2. Typical wall patterns.

only the horizontal joints are accentuated. In the latter treatment the vertical joints, after tooling, are filled with mortar and then rubbed flush, after the mortar has partially hardened, to give it a texture similar to that of the units. In this treatment, the tooled horizontal joints stand out in strong relief. This procedure is well suited to walls or other exposed structural details, where it is desired to emphasize strong horizontal lines. When an especially massive effect is desired, every second or third horizontal joint may be tooled, with the other joints rubbed flush after refilling with mortar.

Another form of joint is obtained by using an excess of mortar when laying the units. Some of the mortar is squeezed out as the units are set and pressed into place. This mortar is not trimmed off but is left to harden. It should be understood that such treatment may produce joints which are not entirely waterproof.

Split units afford another variation in finish. Such units are made by splitting ordinary units lengthwise. The units are laid with their split surfaces exposed. When making the units, many interesting variations can be obtained by introducing mineral colors and by using aggregate of different colors and sizes.

SINGLE-BLOCK AND CAVITY WALLS

Two types of concrete-block walls are commonly used. The most popular of the two is laid up using the kind of units shown at *A, B,* and *C* in Figure 1. Such walls are known as *single-block,* or *solid,* walls. The term *solid* is employed, even though the blocks have from two to three cores in them. This type of wall is illustrated in the following chapters by such illustrations as Figure 14 in Chapter 9, Typical Concrete-block Details, and Figure 12 in Chapter 10, Concrete-block Walls, Foundations, and Pilasters. A less-used but effective wall, especially in terms of insulation, is the *cavity* wall, shown in Figure 3.

A cavity wall consists of two separate walls which are separated by a continuous air space and tied together by means of metal ties. The ties are embedded in the horizontal mortar joints. Neither the inner nor outer walls should be less than 4 inches thick (nominal thickness), and the

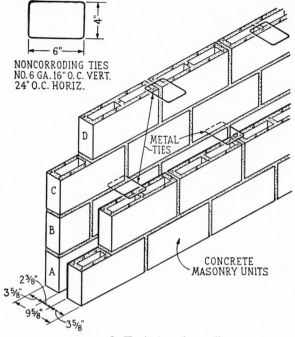

NONCORRODING TIES
NO. 6 GA. 16" O. C. VERT.
24" O.C. HORIZ.

METAL TIES

CONCRETE
MASONRY UNITS

FIGURE 3. Typical cavity wall.

space between them should not be less than 2 inches. Both of the 4-inch walls are laid up using the kind of block shown at *J* in Figure 1.

ALLOWABLE THICKNESSES FOR CONCRETE-BLOCK WALLS ABOVE GRADE AND FOR FOUNDATIONS BELOW GRADE

City and town building codes, especially in earthquake regions, contain strict regulations about the thicknesses and heights of walls and foundations. The following allowable thicknesses are expressed in nominal dimensions and are of a general nature:

Walls for One-story Residences. Units having a thickness of at least 6 inches should be used. If wall height exceeds 9 feet or height to the peak of a gable exceeds 15 feet, wall thickness should be at least 8 inches.

Walls for Two-story Residences. Units having a thickness of at least 8 inches should be used when the total wall height is not more than 35 feet and when the wall is not subjected to lateral thrust from the roof construction.

Foundations for One- and Two-story Residences. Units having a thickness of 8 inches for one-story, and units having a thickness of 12 inches for two-story, houses may be used.

Walls for One-story Nonresidential Buildings. Units having a thickness of 12 inches should be used. However, if the walls are not more than 12 feet high, not over 35 feet long, and if the roof beams are horizontal, units 8 inches thick may be used.

Walls for Two-story Nonresidential Buildings. Units having a thickness of 12 inches should be used. However, if the second story is not more than 12 feet high, 8-inch units may be used.

Foundations for One- and Two-story Nonresidential Buildings. Units having a thickness of at least 12 inches should be used. However, if the foundations for a one-story building do not extend more than 4 feet into the ground, units 8 inches thick may be used.

Cavity Walls for One- and Two-story Residences. The walls should be 10 inches thick and be confined to heights of no more than 25 feet.

Foundations for Cavity Walls. The foundations should be composed of units (*not* cavity) which are at least 12 inches thick.

LATERAL SUPPORT

Concrete-block walls and foundations must be provided with lateral (from-the-side) support which is at right angles to them. Such support may be obtained by cross walls, or pilasters (pilasters are explained in Chapter 10). The supports should not be at greater distances, or intervals, than 20 times the nominal wall or foundation thickness for solid construction and 18 times the nominal thickness where cavity walls or walls of hollow units are concerned.

MODULAR PLANNING

Figure 4 shows several details of what is known as *modular planning*. The module lines, as shown in the plan view part of the illustration, run horizontally and vertically and are 8 inches apart. In other words, each of the squares formed by the horizontal and vertical lines is exactly 8 inches square. The horizontal and vertical lines constitute what is called a *grid*. By the use of such a grid, the planning of concrete-block walls and foundations can be done so as to avoid any necessity for cutting blocks on the job. Walls and foundations planned by the modular method are neat and uniform in appearance. They are also easier to lay up and thus cut down the labor time.

Note the 3′ 4″ window opening in Figure 4. In other words, this opening is 40 inches wide. The width is equal to 2½ of the 16-inch-long blocks. No cutting of blocks would be needed in order to fit them into the necessary space and around the window opening. Note, also, the 2′ 8″ dimension. This distance is equal to 32 inches or the exact length of two of the 16-inch blocks or four half blocks.

The over-all dimension of 20′ 0″ is equal to exactly fifteen of the 16-inch blocks with ⅜-inch mortar joints. Thus, the wall, or foundation, can be laid up using both full and half blocks without any cutting or other difficulties.

The circled details in Figure 4, which indicate a corner and one side of a window opening, show how blocks are laid in walls or foundations and how the actual block dimensions fit into nominal dimensions.

Lengths of Walls and Foundations. When blocks are placed with their long dimension horizontal, they are called *stretchers*. Table 1 shows the nominal lengths of walls or foundations in terms of stretcher units which are 15⅝ and 11⅝ inches long. For example, a wall 20′ 0″ long contains exactly fifteen of the 15⅝-inch stretchers. When planning over-all hori-

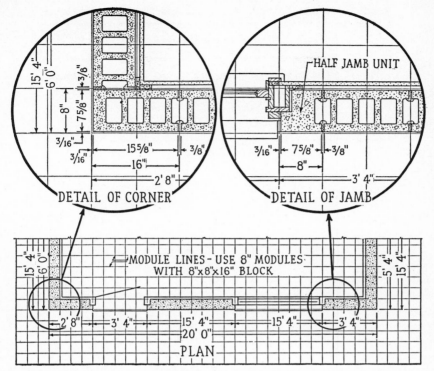

FIGURE 4. Showing how concrete-block walls are dimensioned on drawings for modular planning, using nominal dimensions. Enlarged details show how blocks are actually laid in wall.

zontal dimensions, Table 1 should be consulted and the dimensions controlled to the extent that whole or half units can be used without cutting.

Heights of Walls and Foundations. Each row of units, such as at *A, B, C,* and *D* in Figure 3, is known as a *course*. Table 2 shows the nominal heights of walls and foundations by courses. For example, when units which are 7⅝ inches high are used, nine courses produce a height of 6′ 0″. The over-all heights of all walls and foundations should be planned so that an exact number of courses can be laid without the necessity for cutting blocks.

Openings in Block Walls and Foundations. Any opening, such as for a window, should be carefully planned so that its width and height can be

TABLE 1. Nominal length of concrete-block walls by stretchers

No. of stretchers	Nominal length of concrete-block walls*	
	Units 15⅝ inches long and half units 7⅝ inches long with ⅜-inch-thick head joints	Units 11⅝ inches long and half units 5⅝ inches long with ⅜-inch-thick head joints
1	1' 4''	1' 0''
1½	2' 0''	1' 6''
2	2' 8''	2' 0''
2½	3' 4''	2' 6''
3	4' 0''	3' 0''
3½	4' 8''	3' 6''
4	5' 4''	4' 0''
4½	6' 0''	4' 6''
5	6' 8''	5' 0''
5½	7' 4''	5' 6''
6	8' 0''	6' 0''
6½	8' 8''	6' 6''
7	9' 4''	7' 0''
7½	10' 0''	7' 6''
8	10' 8''	8' 0''
8½	11' 4''	8' 6''
9	12' 0''	9' 0''
9½	12' 8''	9' 6''
10	13' 4''	10' 0''
10½	14' 0''	10' 6''
11	14' 8''	11' 0''
11½	15' 4''	11' 6''
12	16' 0''	12' 0''
12½	16' 8''	12' 6''
13	17' 4''	13' 0''
13½	18' 0''	13' 6''
14	18' 8''	14' 0''
14½	19' 4''	14' 6''
15	20' 0''	15' 0''
20	26' 8''	20' 0''

* Actual length of wall is measured from outside edge to outside edge of units and is equal to the nominal length minus ⅜ inch (one mortar joint).

SOURCE: Courtesy of Portland Cement Association, Chicago, Ill.

TABLE **2.** Nominal height of concrete-block walls by courses

No. of courses	Nominal height of concrete-block walls*	
	Units 7⅝ inches high and ⅜-inch-thick bed joint	Units 3⅝ inches high and ⅜-inch-thick bed joint
1	8″	4″
2	1′ 4″	8″
3	2′ 0″	1′ 0″
4	2′ 8″	1′ 4″
5	3′ 4″	1′ 8″
6	4′ 0″	2′ 0″
7	4′ 8″	2′ 4″
8	5′ 4″	2′ 8″
9	6′ 0″	3′ 0″
10	6′ 8″	3′ 4″
15	10′ 0″	5′ 0″
20	13′ 4″	6′ 8″
25	16′ 8″	8′ 4″
30	20′ 0″	10′ 0″
35	23′ 4″	11′ 8″
40	26′ 8″	13′ 4″
45	30′ 0″	15′ 0″
50	33′ 4″	16′ 8″

* For concrete masonry units 7⅝ inches and 3⅝ inches in height laid with ⅜-inch mortar joints. Height is measured from center to center of mortar joints.

SOURCE: Courtesy of Portland Cement Association, Chicago, Ill.

fitted into a wall or foundation without having to cut blocks. Table 3 shows the modular sizes of window openings which can be used in walls or foundations to avoid the necessity for cutting blocks.

Illustrative Example. Figure 5 shows a typical block wall with one window and one door opening. Nominal block size is 8 by 8 by 16 inches.

The 17′ 4″ over-all length of the wall is correctly planned so that it contains exactly thirteen full stretchers. Or it could contain eleven full stretchers and two half stretchers. Thus the wall can be laid up without the necessity for cutting blocks. If the over-all dimension had been 17′ 0″,

TABLE 3. Modular concrete-block openings for wood window frames

Types of windows	Masonry openings*	Glass size	Masonry openings	Glass Size	Masonry openings	Glass size
Double	2′ 0″ × 3′ 4″	16″ × 12″	2′ 8″ × 3′ 4″	24″ × 12″	3′ 4″ × 3′ 4″	32″ × 12″
	4′ 0″	16″	4′ 0″	16″	4′ 0″	16″
	4′ 8″	20″	4′ 8″	20″	4′ 8″	20″
	5′ 4″	24″	5′ 4″	24″	5′ 4″	24″
	6′ 0″	28″	6′ 0″	28″	6′ 0″	28″
	6′ 8″	32″	6′ 8″	32″	6′ 8″	32″
Double	4′ 0″ × 4′ 0″	40″ × 16″	4′ 8″ × 4′ 0″	48″ × 16″	5′ 4″ × 4′ 0″	56″ × 16″
	4′ 8″	20″	4′ 8″	20″	4′ 8″	20″
	5′ 4″	24″	5′ 4″	24″	5′ 4″	24″
	6′ 0″	28″	6′ 0″	28″	6′ 0″	28″
	6′ 8″	32″	6′ 8″	32″	6′ 8″	32″
	7′ 4″	36″	7′ 4″	36″	7′ 4″	36″
Casement	1′ 4″ × 3′ 4″	8″ × 25″	2′ 0″ × 3′ 4″	16″ × 25″	2′ 8″ × 3′ 4″	24″ × 25″
	4′ 0″	33″	4′ 0″	33″	4′ 0″	33″
	4′ 8″	41″	4′ 8″	41″	4′ 8″	41″
	5′ 4″	49″	5′ 4″	49″	5′ 4″	49″
	6′ 0″	57″	6′ 0″	57″	6′ 0″	57″
	6′ 8″	65″	6′ 8″	65″	6′ 8″	65″
Basement	Two light sash		Three light sash			
	2′ 0″ × 2′ 0″	8″ × 12″	2′ 8″ × 2′ 0″	8″ × 12″		
	2′ 8″	20″	2′ 8″	20″		
	2′ 8″ × 2′ 0″	12″ × 12″				
	2′ 8″	20″				

NOTE: Modular masonry openings shown above should also be used for metal window frames. It may be necessary, however, to provide metal surrounds to fit the metal frames into the modular openings.

* Masonry openings are dimensioned from jamb to jamb and from bottom of lintel to bottom of precast concrete sill. Openings of sizes other than those shown may be used by keeping the dimensions in multiples of 8 inches. Glass sizes shown are for one light, but any division of the sash may be used.

the wall would have ended as shown at X and would have necessitated the cutting of blocks.

The window size is not correctly planned so that the required opening can be laid up without cutting blocks. The 3′ 8″ width equals 44 inches which is not a multiple of 16 inches. In other words, the width is 4 inches too narrow to equal exactly the length of four of the full-length stretchers plus their ⅜-inch mortar joints. Thus, as shown by the shaded blocks on both sides of the window opening, blocks would have to be cut. This adds unnecessary labor and material costs and ruins the appearance of the wall. The 5′ 0″ height of the window opening includes more than seven courses and fewer than eight courses. This means that blocks in one course have to be cut, as shown by the shading. The location of window openings must also be planned so that either full or half stretchers can be used between them and the corners of the wall.

The door size is not correctly planned so that the necessary opening can be laid up without cutting blocks. The 3′ 2″ dimension equals 38 inches which is not a multiple of 16 inches. In other words, the width is 6 inches

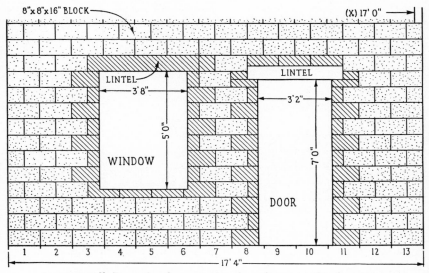

FIGURE 5. Example of wrong planning of concrete-block wall.

too wide for two full stretchers and 10 inches too narrow for three full stretchers. The shaded blocks on both sides of the door are those which would have to be cut. The 7′ 0″ height of the door opening includes more than ten courses and fewer than eleven courses. Thus, blocks have to be cut, as indicated by the shading.

The lintels over window and door openings must also be considered when planning openings. In general, masonry lintels, as shown in Chapter 9, Typical Concrete-block Details, should be of the same height as the blocks and have a bearing (amount of space required for support) equal to half a block at both ends.

Figure 6 shows how the window and door openings of Figure 5 should have been planned in order to avoid the necessity for cutting blocks, in order to bring about economy in labor and materials and in order to create a uniform and pleasing appearance. Note that only the block at *Y* varies from the running-bond pattern but that it is a half length which can be laid up without cutting.

Before planning block walls, it is recommended that information be obtained about the exact sizes of blocks available. The modular sizes, as

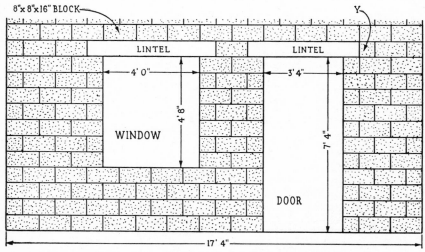

FIGURE 6. Example of correct planning of concrete-block wall.

herein exhibited, are also recommended, along with modular-sized windows and doors.

BONDING

Previously, the term *bond* was used to indicate the adhesiveness of mortar to block. In other words, the term was used in the sense of joining or binding concrete blocks together. The term *bonding,* as indicated in Figure 2, concerns the arrangement of units in a wall or foundation so that their overlapping thoroughly and securely ties the units together, enabling the whole wall or foundation to act as one piece in resisting wind pressure and other stresses. Some types of bonding, such as vertical and horizontal stacking, do not contain overlapped units, and there is some question as to the advisability of their use in walls and foundations which must support loads and resist other stresses.

JOINTS AND TOOLING

The term *joint* means the mortar between units. The term *tooling* refers to the process of compressing and forming the mortar joints to produce desired shapes and appearances.

There is evidence to prove that relatively thin joints produce stronger walls and foundations than thick joints. For example, ⅜-inch joints are

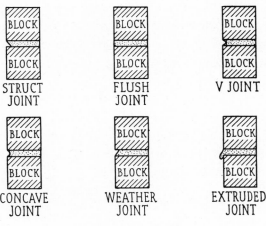

FIGURE 7. Joints.

recommended for all structural work instead of the ½-inch joints which were popular some time ago. The generally used kinds of joints are shown in Figure 7 and explained in the following:

Struct Joint. This kind of joint is satisfactory for interior walls even though it is a dust catcher. It is not recommended for an exterior wall or for foundations because of the possibility of its causing leaks. Such a joint is formed by first removing the excess mortar and then running the point of a trowel along it.

Flush Joint. This kind of joint is satisfactory where no special effects are desired. It is made by keeping the trowel parallel to the faces of walls or foundations while drawing the point of the trowel along the length of the joint.

V Joint. For this kind of a joint, a special tool is required. First, a flush joint is made. Then the special tool is pressed into the joint mortar and drawn along the full length of the joint. This tool is repeatedly used until the joint is smooth and well formed.

Concave Joint. This kind of joint is made like a V joint except that a special rounded tool is required.

Weather Joint. This kind of joint is designed to shed water readily. It is made by striking it downward with the top edge of the trowel. This presses or compacts the mortar into a tight joint which is especially desirable where driving rains are the rule.

Extruded Joint. While this type of joint is not weather and moisture-proof, it has a desirable decorative effect and can be used for structural details in which weather-tightness is of no concern. In making such a joint, an excess amount of mortar is spread. The units are pressed into place allowing the mortar to squeeze out.

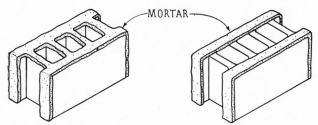

FULL MORTAR BEDDING   FACE-SHELL MORTAR BEDDING

FIGURE 8. Examples of full bedding and face-shell mortar bedding.

Full Mortar Bedding. Figure 8 shows an example of full mortar bedding and of face-shell mortar bedding of a block. For most work, the face-shell mortar bedding is satisfactory.

CUTTING CONCRETE BLOCK

If walls and other structural details are properly planned, blocks will never have to be cut. However, there are instances when parts of block have to be cut for special purposes. For the cutting work, a broad, heavy chisel, such as shown in Figure 9, should be used, along with a heavy hammer. The block should be scored along the cutting line, after which heavier blows are struck until the block splits. With care, cutting can be done with exactness.

Sometimes blocks have to cut so as to provide room for electrical switch boxes or outlets. Such cutting can be done to better advantage by

FIGURE 9. Example of how to cut concrete block.

using a small cold chisel and a hammer. The outline of the hole should first be scored. Then heavier blows can be struck and small pieces of block cut out a few at a time until the hole is large enough for the purpose intended.

CONTROL JOINTS

In long walls and foundations, such as in barns or other large structures, some shrinkage is apt to occur. The shrinkage sets up stresses in the wall which, if not relieved, are apt to cause cracks, especially around window and door openings. To relieve such stresses and avoid cracks, control joints are desirable.

Such relief can be brought about if a means is provided by which walls and foundations can move a little (shrink or expand) without endangering alignment or stiffness. There are many ways of accomplishing such a purpose. The following method is a typical example:

The upper part of Figure 10 shows whole- and half-length control joint blocks at A and B and at C and D. Note that blocks A and D have tongues and that blocks B and C have grooves for insertion of tongues.

The lower part of Figure 10 shows a portion of a wall in which such control blocks are employed to create a vertical control joint. This part of the illustration shows how the tongue of block D fits into the groove of block C. Blocks A and B fit together the same way. Alternate courses use blocks D and C and blocks A and B. The control joint is not mortared.

In order that control joints be weather-tight, they should be filled with an elastic caulking compound.

If a wall, for example, shrinks a little, it can pull slightly apart at the control joint. Such movement is made possible by the tongues sliding along the grooves. The movement, which is not likely to be more than ¼ inch, can take place without endangering the alignment or rigidity. Shrinkage stresses are thus relieved, and there will be no danger of cracks.

Control joints should be laid up at intervals of 25 to 30 feet.

REINFORCED-CONCRETE BLOCK

In Chapter 4, Concrete, it was pointed out that concrete has great compressive strength and that it will easily and safely support enormous loads so long as it, in turn, is soundly supported. Earlier in this chapter it was pointed out that walls, made of concrete block, for example, also have all the compressive strength required for most purposes. Such strength, as with placed concrete masses, depends upon the blocks being soundly supported.

There are several observations on concrete-block construction pointed

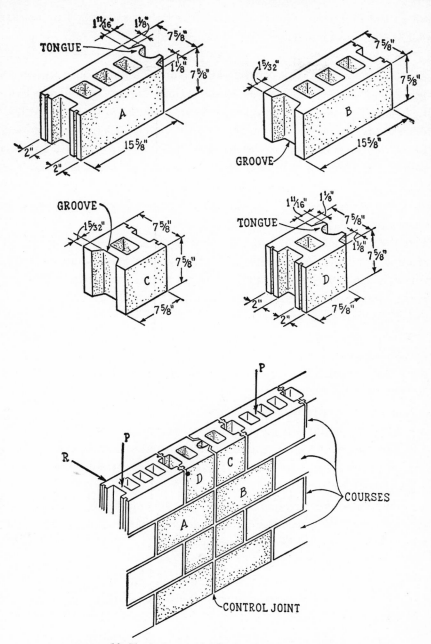

FIGURE 10. Typical control-joint block and control joint.

out in the following. They should be thoroughly understood by anyone planning or erecting concrete-block structures.

Load Application. If a concrete-block wall or foundation is supported by ample footings (footings are explained in Chapter 5, Concrete Footings) and is carefully and properly laid up, using good mortar, it will be able to support very heavy weights or loads safely so long as those loads are applied as indicated by the arrows marked P in Figure 10. In other words, if the loads are applied (as from floors and roofs) in a straight-down direction, they can be easily and safely supported. Then, the loads create a compressive stress which concrete-block structures can resist or support to great advantage.

If the loads are applied in the direction of the arrow marked R, for example, by wind or a sloping rafter, a different stress would be set up which is somewhat similar to the tensile stresses described in Chapter 4, Concrete. Such stresses would tend to tear a wall apart by separating the blocks from their mortar joints.

If the footing supporting a block structure, such as a foundation, were not sufficiently strong and allowed the foundation to settle here and there, the stresses set up would tend to cause cracks and separate the blocks at the joints.

In long, high, exposed walls, severe wind pressure could set up other stresses which would tend to tear such walls apart at the joints.

When concrete blocks are used to support a bank of soil, the weight of the soil causes a load which is applied in a direction similar to the arrow marked R in Figure 10.

In regions where earthquakes occur, the movement caused in walls and foundations tends to tear them apart and cause large cracks at the joints. Such damage tends to destroy strength.

Openings in Walls. Over window and door openings, the blocks and mortar joints, alone, cannot supply the tensile strength necessary to resist the stresses thus created by the weight of the blocks, plus other possible loads, over the openings. There must be some form of lintel (lintels are explained in Chapter 9) or other means provided for supporting the blocks over such openings.

From the foregoing, it is evident that in many cases concrete-block structures must have some means of additional strength built into them. Such strength, or resistance to unusual stresses, can be provided by means of steel reinforcement.

Typical Steel Reinforcement. It should be pointed out that most cities and towns have rigid codes, especially in regions where high winds or earthquakes are likely, which specify the amount, kind, size, method of application, and other details of reinforcement. Readers are urged to check their local codes. The following suggestions are merely typical and constitute a means of showing several ways in which reinforcement can be used.

Figure 11 shows typical methods of reinforcing concrete-block walls. The same methods apply to foundations.

The bond beam, shown at *A*, is laid up, using a block similar to the block shown at *K* in Figure 1. Such beams are constructed by the use of two or more steel rods (generally about ½ inch in diameter) and concrete made of small-pebble gravel or crushed stone. When the concrete hardens, the beam acts as a solid member, and the steel provides the necessary tensile strength. Beams of this kind are used over openings, such as the door in Figure 11, and around the tops of walls and foundations as a means of tying all blocks together and distributing the loads to better advantage.

Many building codes specify that reinforcement, called *studs,* be used at all corners, on both sides of openings, at intervals along the lengths of walls and foundations, and at intersections where walls meet walls or foundations meet foundations. The studs are extended down to the footings. All cores in which studs are placed are filled with the small-aggregate concrete previously mentioned. To all intent and purposes, the studs act as reinforced-concrete columns and provide the strength to resist unusual stresses. See *B* in Figure 11.

If the soil under footings is not reliably dense and strong, as explained in Chapter 5, Concrete Footings, two or more steel rods can be used, as shown at *C* in Figure 11.

Other effective reinforcement can be accomplished by placing small steel rods in the horizontal mortar joints, as shown at *D* in Figure 11.

Any reinforcing is only as good or effective as the care used in its installation and the quality of the mortar or concrete used in conjunction with it. Reinforced-block walls and foundations are excellent and dependable if properly constructed. Too much emphasis cannot be placed

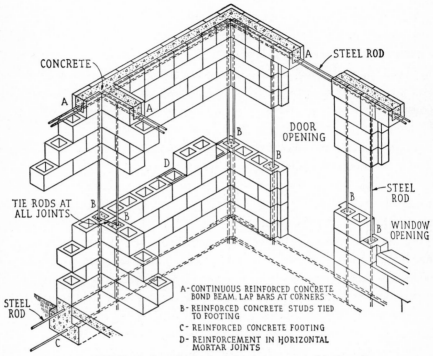

FIGURE 11. Typical method of reinforcing concrete-block walls.

upon the necessity for careful proportioning of mortar and concrete mixes and upon workmanship. When these qualities are good, the resulting structural work will be durable.

TERMITE CONTROL

In some regions termites constitute a threat, and careful precautions should be taken in building the intersections of walls and floor or foundations and floors.

When concrete floor slabs are placed directly on ground, the usual practice has been to make an expansion joint, or moisture joint, between the edge of the floor and the wall or foundation. Such joints are frequently filled with tar, asphalt, or wood. Termites can and do eat their way through such joints and thus gain access to wood details.

Figure 12 shows a method whereby there is no joint through which termites can gain access to wood details. The method consists of using a

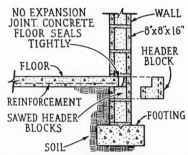

FIGURE 12. Example of how to prevent termite entry.

TABLE 4. Quantities of materials for concrete-block walls

Actual unit sizes (width × height × length), inches	For 100 square feet of wall			For 100 Units
	Nominal wall thickness, inches	Number of units	Mortar* cubic feet	Mortar* cubic feet
$3\frac{5}{8} \times 3\frac{5}{8} \times 15\frac{5}{8}$	4	225	4.3	1.9
$5\frac{5}{8} \times 3\frac{5}{8} \times 15\frac{5}{8}$	6	225	4.3	1.9
$7\frac{5}{8} \times 3\frac{5}{8} \times 15\frac{5}{8}$	8	225	4.3	1.9
$3\frac{3}{4} \times 5 \times 11\frac{3}{4}$	4	221	3.7	1.7
$5\frac{3}{4} \times 5 \times 11\frac{3}{4}$	6	221	3.7	1.7
$7\frac{3}{4} \times 5 \times 11\frac{3}{4}$	8	221	3.7	1.7
$3\frac{5}{8} \times 7\frac{5}{8} \times 15\frac{5}{8}$	4	112.5	2.6	2.3
$5\frac{5}{8} \times 7\frac{5}{8} \times 15\frac{5}{8}$	6	112.5	2.6	2.3
$7\frac{5}{8} \times 7\frac{5}{8} \times 15\frac{5}{8}$	8	112.5	2.6	2.3
$11\frac{5}{8} \times 7\frac{5}{8} \times 15\frac{5}{8}$	12	112.5	2.6	2.3

Table based on $\frac{3}{8}$-inch mortar joints.

* With face-shell mortar bedding. Mortar quantities include 10 per cent allowance for waste.

special header block, such as shown at *F* in Figure 1, so that the concrete floor can rest on the header. Such headers must have a greater depth than the 2¾ inches shown in Figure 1. Once the course of headers is laid up and the floor placed, the next regular course of stretchers can be laid in the usual manner. Then, there is no joint through which termites can gain access to wood details above.

ESTIMATING

Table 4 shows the quantities of various-sized blocks and the mortar necessary for 100 square feet of concrete-block wall or foundation. For areas greater or less than 100 square feet, the table quantities can be divided or multiplied. For example, 50 square feet of wall would require only half the quantities shown in the table.

CHAPTER NINE

Typical Concrete-block Details

When the term *detail* is used in connection with concrete block, it generally has two meanings. First, it may mean a planning process wherein certain aspects of walls and other structural parts of buildings are carefully considered and planned prior to the time construction work is actually started. For example, an architect prepares the general plans for a proposed house and indicates the walls, roofs, supporting members, and the dimensions required. He also prepares special plans which show the intersection of walls and floors, how the roof is connected to the exterior walls, how loads are to be supported, etc. Such special planning is known as *detailing*. In other words, it is the process of planning or figuring out how various parts of a structure are to be assembled or built. Second, the term *detail* is used to indicate a special plan after it has been drawn and made ready for use by mechanics. For example, the wall over a window or door opening has to be supported. When that support has been planned and indicated in a special drawing, it is known as a *detail drawing*.

The general planning for any structural work must be accompanied by details. Unless the details are carefully planned before structural work gets under way, the actual construction work will be subject to an endless list of troubles and difficulties. This is especially true of unit masonry, such as concrete block, where pieces of given size and shape must be used without cutting and so as to serve the structural needs.

The purpose of this chapter is to illustrate and explain some of the commonly encountered details of concrete-block planning and construction.

161

These data are presented to assist readers in planning and constructing small houses and other projects. In all construction work, safety, durability, and economy are of supreme importance. This chapter gives the fundamentals of such good construction. The details shown are typical and are intended to be helpful in preparing plans. However, readers are urged to look up their local building codes and to make plans accordingly.

In all the details illustrated and explained in this chapter, modular planning, because of the resulting economy, is assumed. Therefore, it should be kept in mind that window and door frames, for example, must also be modular in dimensions. Such dimensions can be obtained from most mills.

LINTELS

When window openings, such as illustrated in Plate II, and door openings are required in concrete foundations or walls, some provision must be made to support the weight of the blocks and other possible loads, above the openings. In the following explanations, only window openings are considered. However, exactly the same procedure applies equally well to door openings. The means of supporting the weights and loads above such openings are known as *lintels*.

The A part of Figure 1 shows a typical window opening in an ordinary concrete-block wall. The width of the opening, from x to y, is the length (span) of the block wall which must be supported. The P_1 arrow indicates how the weight of the blocks above the opening acts downward. Unless there is a lintel over the opening, the wall above would not stay in place. The shaded area indicates the lintel and shows that each end of it must have a bearing surface equal in length to a half block, or about 8 inches.

The section view at B in Figure 1 shows the same window opening. Note that the lintel is of the same thickness as the block. The P_1 arrow indicates the weight of the block above the opening. In this case, the lintel supports only the weight of the block above the opening.

In many instances the ends of floor joists have to be supported by walls or foundations. Sometimes, as indicated in the section view at C in Figure

1, such joists must be supported by walls directly over window openings. Then the lintel has to support not only the weight of the blocks above it but also some of the floor load. The P_1 and P_2 arrows indicate the weight and the load.

How to Calculate Weight and Load. Suppose, for example, that the wall or foundation, shown at *A* in Figure 1, extends a distance of 10′ 0″

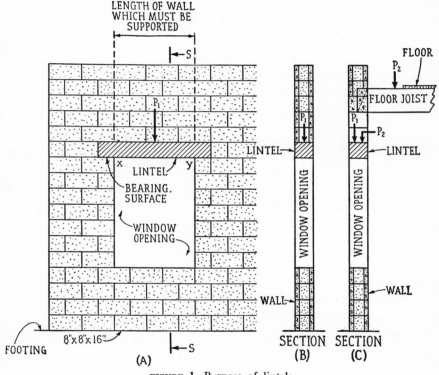

FIGURE 1. Purpose of lintels.

(exactly 15 courses—see Table 2 in Chapter 8, Concrete Block), above the top of the lintel. Further suppose that the window opening is 3′ 4″ wide. This is exactly two and one-half stretchers. Ordinary blocks weigh about 60 pounds per cubic foot. Since 8 inches (nominal thickness of the block) equals ⅔ foot, the volume of this section of the wall or foundation is ⅔ × 3⅓ × 10, or approximately 22 cubic feet. The weight of the section is therefore 22 × 60, or 1,320 pounds. This is the weight of the blocks which the lintel must support.

Next suppose that, as shown in the section view at *C*, the wall or foundation has to support part of a floor load. This situation is indicated in Figure 2. The joists which support the floor are in turn supported by the wall or foundation at one end and by a girder at the other end. In other words, the floor load is equally divided between the wall or foundation and the girder. The width of the window opening (as shown in the *A* part of Figure 1) is equal to two and one-half of the stretchers (horizontal position of the block), or 3′ 4″. Half of the floor span is 9′ 0″. The floor

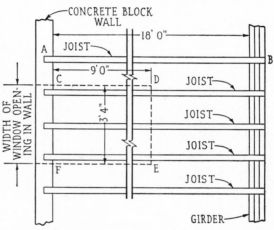

FIGURE 2. How lintels support floor loads.

area supported by the wall or foundation is therefore 3⅓ × 9, or 30 square feet. If the combined dead and live load for the floor is 70 pounds per square foot, the total load is 70 × 30, or 2,100 pounds. Adding this to the weight of the block over the opening produces 1,320 + 2,100, or 3,420 pounds which the lintel must be able to support.

The weight of 1,320 pounds and the load of 2,100 pounds are both appreciable, and it can be seen that lintels must be carefully selected and laid. The foregoing example was given merely to bring out the fact that lintels are important. Fortunately, there is an easier way of selecting them for most houses and other small buildings.

Kinds of Lintels. There are several varieties of concrete lintels which are frequently used in connection with concrete-block walls and foundations. All such lintels, in order to provide the safety factor desirable, should have steel reinforcing rods in them. The steel, as previously pointed out, increases the tensile strength of concrete.

FIGURE 3 (*A*). This is the type of lintel which is most commonly employed over openings in ordinary walls and foundations. The recess at *x* is to accommodate the head or top parts of some types of windows and doors. However, lintels can be obtained without such a recess.

Table 1 shows the number and size of steel bars which must be used in lintels over various spans (such as *xy* in the *A* part of Figure 1) when only the weight of the blocks above the openings has to be supported. For ex-

TABLE 1. Lintels with wall or foundation weight only

Size of lintel		Clear span of lintel, feet	Bottom reinforcement	
Height, inches	Width, inches		No. of bars	Size of bars
5¾	7⅝	Up to 7	2	⅜-inch round deformed
5¾	7⅝	7 to 8	2	⅝-inch round deformed
7⅝	7⅝	Up to 8	2	⅜-inch round deformed
7⅝	7⅝	8 to 9	2	½-inch round deformed
7⅝	7⅝	9 to 10	2	⅝-inch round deformed

ample, the lintel shown in Figure 1 must have two steel bars, each of which is ⅜ inch in diameter.

NOTE: The term *deformed* means that the rods have rough surfaces which prevent them from slipping in the concrete.

FIGURE 3 (*B*). This type of lintel, sometimes known as a *split* lintel, may be desirable in some circumstances. It can be secured with or without the recess shown at *x*.

Table 2 shows the number and size of steel bars which must be used in two-piece lintels over various spans when only the weight of the blocks above the openings has to be considered. For example, if a two-piece lintel is to be used over an opening, such as shown in the *A* part of Figure 1, each piece must have one bar which has a diameter of ⅜ inch.

FIGURE 3 (*C*). This type of lintel, which has additional bars in the form of stirrups (see the dashed lines in the section view) and two additional

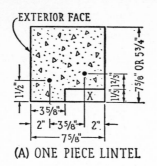

(A) ONE PIECE LINTEL

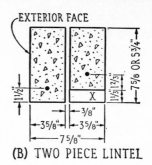

(B) TWO PIECE LINTEL

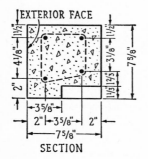

SECTION

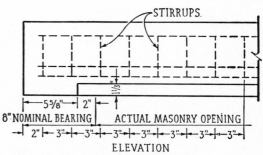

ELEVATION

(C) ONE PIECE LINTEL WITH STIRRUPS

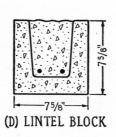

(D) LINTEL BLOCK

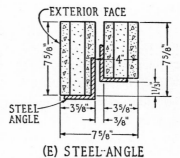

(E) STEEL-ANGLE

FIGURE 3. Lintel reinforcing.

horizontal bars (see the dashed lines in the elevation view) is used when a lintel must support both the weight of blocks above an opening and the load from a floor. It, too, can be secured with or without the recess. NOTE: Two-piece lintels should not be used to support combined wall or foundation weight and floor load because it is difficult to design the inner piece to have the same strength as the outer piece. Differences in strength between the two pieces would result in cracks in walls or foundations.

TABLE 2. Two-piece lintels with wall or foundation weight only

Size of lintel		Clear span of lintel, feet	Bottom reinforcement	
Height, inches	Width, inches		No. of bars	Size of bars
5¾	3⅝	Up to 7	1	⅜-inch round deformed
5¾	3⅝	7 to 8	1	⅝-inch round deformed
7⅝	3⅝	Up to 8	1	⅜-inch round deformed
7⅝	3⅝	8 to 9	1	½-inch round deformed
7⅝	3⅝	9 to 10	1	⅝-inch round deformed

Table 3 shows the number of horizontal rods and stirrups which must be used over various spans when *both* the weight of the block above the openings and the floor load have to be considered. For example, the lintel shown in the section at *C* in Figure 1 would require two top bars, each ⅜ inch in diameter, two bottom bars, each ¾ inch in diameter, and three stirrups spaced as shown. If the span were 5 feet, four bars again would be necessary. The top two would have to be ⅜ inch in diameter and the bottom two ⅞ inch in diameter. Five stirrups would be necessary.

FIGURE 3 (*D*). This is the same general type as shown at the top of the wall illustration in Figure 11 of Chapter 8, Concrete Block. It is used only when it is a part of a bond beam which extends all around a wall or foundation. Bond beams are further explained in succeeding pages.

FIGURE 3 (*E*). The lintels shown in the *A*, *B*, *C*, and *D* parts of this illustration are all made of solid concrete without cores. On the other hand, the blocks shown in the *E* part of the illustration have cores and are similar to the block shown at *J* in Figure 1 of Chapter 8, Concrete Block. When such blocks are used as lintels, they must be supported by steel

TABLE 3. Lintels with wall or foundation weight and floor loads

| Size of lintel | | Clear span of lintel, feet | Reinforcement | | Web reinforcement No. 6-gauge wire stirrups. Spacings from end of lintel—both ends the same |
Height, inches	Width, inches		Top	Bottom	
7⅝	7⅝	3	None	2–½-inch round	No stirrups required
7⅝	7⅝	4	2–⅜-inch round	2–¾-inch round	3 stirrups, Sp.: 2, 3, 3 inch
7⅝	7⅝	5	2–⅜-inch round	2–⅞-inch round	5 stirrups, Sp.: 2, 3, 3, 3, 3 inch
7⅝	7⅝	6	2–½-inch round	2–⅞-inch round	6 stirrups, Sp.: 2, 3, 3, 3, 3, 3 inch
7⅝	7⅝	7	2–1-inch round	2–1-inch round	9 stirrups, Sp.: 2, 2, 3, 3, 3, 3, 3, 3, 3 inch

NOTE: The floor load is assumed to be 85 pounds per square foot with a 20-foot span.

angles. For all ordinary small buildings, the steel angles may have legs which are 3 or 4 inches wide.

The foregoing information and data relative to lintel types and reinforcing can safely be used in all ordinary and commonly encountered small buildings, such as houses, stores, barns, and garages. For larger buildings, especially where greater than usual weights and loads are involved, a structural engineer should be consulted.

Cast-in-place Lintels. If precast concrete lintels are not available, suitable substitutes can be cast in place as a concrete-block wall or foundation

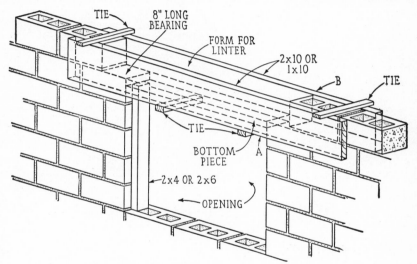

FIGURE 4. Form for cast-in-place lintel.

is being laid. Figure 4 shows a typical opening, such as for a window, and a suggested design for the necessary forms.

The side pieces *A* and *B* of the form can be made of 2 by 10s each of which is about 4 feet longer than the opening is wide. The bottom piece can be made from a 2 by 8 whose length is exactly equal to the width of the opening. The side pieces should be securely nailed to the bottom piece. Ties can be used to prevent the forms from spreading and to keep the 2 by 10s snugly against the block. Care should be exercised in making all joints as tight as possible to prevent leakage of cement paste.

Following the data given in Figure 3 and in Tables 1 and 3, steel reinforcement should be buried in the concrete as it is being placed.

Concrete for lintels can be made with a $1:2\frac{1}{4}:3$ mix and 6 gallons of water per bag of cement, as recommended in Table 1 of Chapter 4, Concrete.

How to Make Precast Lintels. The lintels required for all openings in concrete-block walls or foundations can be precast at a job site before the laying of block is started. The *A* part of Figure 5 shows typical forms and a platform.

The sides and ends of the form can be made using 2 by 8s if their widths are, or can be planed down to, exactly $7\frac{5}{8}$ inches. The end pieces

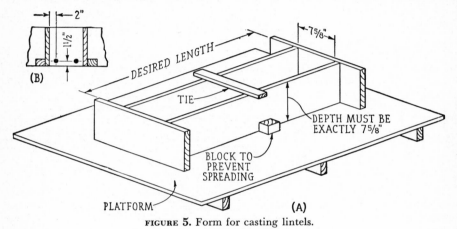

FIGURE 5. Form for casting lintels.

should be securely nailed to the side pieces. In addition, a tie and two or more blocks should be used to keep the forms from spreading. The platform should be flat and level.

Following the data given in Figure 3 and in Tables 1 and 3, steel reinforcing rods should be buried in the concrete. The *B* part of Figure 5 shows the use of two rods. Wire hangers can be obtained from masonry dealers and used to keep the rods in position as the concrete is being placed.

WINDOW SILLS

Both metal and wood window frames, like everything else in structural work, must be supported in such a manner that they cannot sag, tilt, or crack. In concrete-block walls and foundations, such support is provided by means of concrete *sills*. The sills span the full widths of windows and are in turn supported by the blocks under them. Such sills are all in one integral piece and thus provide a stiff and strong base for window frames, and at the same time, add finish to the lower part of such openings. They also prevent the entry of water, wind, and dirt.

Precast Sills. Figure 6 shows a precast window sill in place and a window frame being set into position over it. Note that the sill has a *wash*

FIGURE 6. Precast concrete window sill in place under a frame. (Courtesy of Portland Cement Association.)

(slope) on the exterior side so that water will drain off quickly and completely. The sills are made to project about 1 inch beyond the surfaces of walls and foundations and are provided with drips on the lower and outer edges so that water running off them will fall free rather than run down surfaces, possibly leaving unsightly stains.

Figure 7 shows the details of concrete sills to be used with metal and wood window frames. The top views can be visualized as though the sills are viewed from positions directly above them.

Precast sills are sometimes installed after the walls or foundations have been laid up. When installed at the time the walls or foundations are being laid up, as shown in Figure 6, they must be protected against possible breakage or staining during construction. The joints under sills

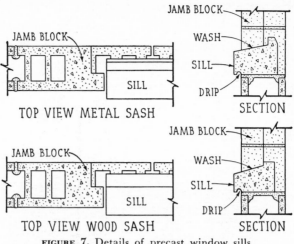

FIGURE 7. Details of precast window sills.

should be completely filled with mortar and carefully tooled. At the ends of the sills, the joints should be filled with mortar or with elastic caulking compound.

Cast-in-place Sills. If precast concrete sills are not available, suitable substitutes can be cast in place after a concrete wall or foundation has been laid up. Figure 8 shows a mason troweling the surface of such a sill and a suggested design for the necessary forms. Note that the window frame is already in place and that the wash of the sill has to be formed by the mason.

Before placing the concrete for such a sill, the cores in the block under the sill must be plugged to prevent the concrete from running down into them. This can be accomplished by the use of heavy screening, metal lath, or wads of paper.

Combination Sills. If precast or cast-in-place sills of the kinds previously explained are not desirable from the standpoint of cost or otherwise, another kind can be built on the job as walls or foundations are being laid up or after all such rough structural work has been completed. This kind of sill, known as a *combination* sill, has the advantage of economy, especially, if labor cost is not of concern, and of being better able to resist earthquake stresses or other unusual pressures.

FIGURE 8. Cast-in-place concrete window sill. (Courtesy of Portland Cement Association.)

The *A* part of Figure 9 shows a picturelike view of a typical window opening in a concrete-block wall such as might be used for a house located in a warm climate. Ordinary 8 by 8 by 16 inch blocks of the two-core variety and square jamb blocks are assumed. The opening is three stretchers wide and five courses high.

The sill is a combination of 3⅝-inch stretchers, such as shown in Figure 1 of Chapter 8, Concrete Block, and cast *pebble* concrete. Pebble concrete is made by using very small pebbles of crushed rock or gravel.

If window frames not having modular dimensions must be used, as is often the case in localities far removed from large mills or building-material dealers, combination sills serve an added economy and purpose. Suppose, for example, that modular blocks having the 7⅝- and 15⅝-inch dimensions were available and that modular precast sills and frames were not available. The following explanations suit such a condition:

The top or head of a window frame for an opening, such as shown in the *A* part of Figure 9, should be at the position of *B* so that the lintel can be placed above it. In other words, the top of the frame should be at the height of the edge *DB* of the block. Further suppose that the frame at hand has a total height, from the top of its head to the bottom of its sill, which is just about 6 inches less than the distance *AB*. The problem is to construct a sill that will make the frame fit the opening.

The 3⅝-inch block shown in the *A* part of the illustration makes a vertical height of 4 inches when bedded down in the usual mortar joint.

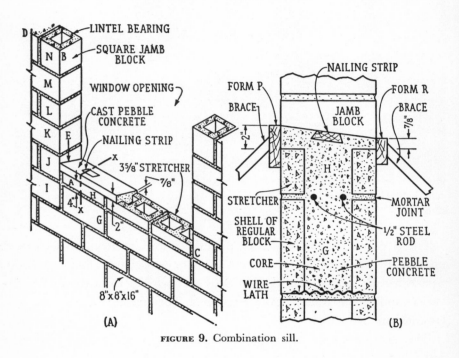

FIGURE 9. Combination sill.

Two more inches have to be built up. It should be remembered that sill portions of a window frame slope down from inside to outside, as shown at *E*. Thus, if the interior of the frame requires 2 inches, the exterior will require only about ⅞ inch. Such dimensions are shown in the illustration.

The additional build-up can be secured by casting pebble concrete as indicated. Three nailing strips should be keyed into the cast concrete so that the wood sill of the frame can be nailed securely to them.

The *B* part of Figure 9 shows an enlarged section view. To visualize the section view it is necessary to imagine that the combination sill can be cut through the line *X-X* as shown in the *A* part of the illustration. It should be understood that the cutting line X-X is so located that it goes through the cores of both the $3\frac{5}{8}$-inch and regular stretcher under the sill. The section view indicated in the B part of Figure 9 shows the $3\frac{5}{8}$-inch block H over a regular stretcher block lettered G. This is also shown in the A part of the illustration. The core of both the H and G block match so as to constitute a continuous and vertical hole or stud. The same is true for other blocks under the window-opening span.

As the $3\frac{5}{8}$-inch blocks are laid, two steel bars should be placed in the joint. If such a sill is made before courses J, K, L, M, and N are laid up, the rods can be extended from 8 to 24 inches into the joint under course J if that joint is made a little thicker than usual. Or the rods can be bent down so as to go into the first cores of the jamb block where vertical rods, as shown in Figure 11 of Chapter 8, Concrete Block, are located. The steel lends a great deal of added strength and stiffness to the sill and the portion of wall under it.

In the mortar joints under course 1 beneath the span of the sill, wire lath should be placed.

The forms for the cast concrete can be made of boards held against the block by braces. The top of form P should be 2 inches above the top of the $3\frac{5}{8}$-inch block. The top of form R should be $\frac{7}{8}$ inch above the top of the $3\frac{5}{8}$-inch block.

Pebble concrete can be made using one part cement, two parts sand, and four or five parts pebbles. Very coarse sand can be used if pebbles are not available. In such a case, the mix can be one part cement and three parts coarse sand. Use just enough water so that the mix will flow easily and can be spaded into the cores and between the forms.

Place the concrete by first filling up the cores. The wire lath will stop the concrete from going deeper than required. Use a piece of steel rod or a small-diameter stick to spade the concrete in the cores. Use a somewhat stiffer mix when placing the sill. Place the nailing strips in the concrete before setting takes place. Smooth the surface of the sill to obtain the necessary wash between forms P and R.

The forms can be removed after the concrete has set and feels hard to the touch. The vertical sides of the new concrete can be rubbed with a trowel until smoothness has been achieved. The window frame should not be nailed into place for at least 4 days.

FASTENING JAMBS TO BLOCK WALLS

In order to make the jambs (side of window and door frames) secure in concrete-block openings, two methods of fastening are commonly employed.

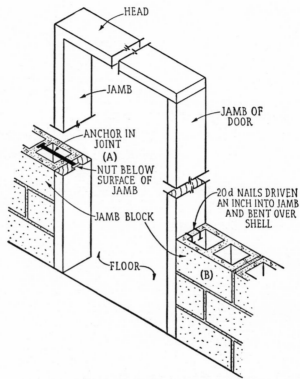

FIGURE 10. How to fasten jambs to block.

Figure 10 shows a typical door frame in a partially laid-up concrete-block wall. The appearance of such frames varies, and the one shown in the illustration is merely an example.

For heavy wood jambs, such as are generally used for stores and other commercial buildings, two or three anchor bolts, as shown at *A*, are used for each side of the jamb. If the frames are set in place before the openings are laid up, holes for the bolts can be bored in the frames and the bolts inserted in the holes and mortar joints as the jamb blocks are laid up. If the jamb blocks are laid up before the frames are set in place, the bolts can be bedded in mortar joints, as shown in the illustration. In any event, the bolts should be so placed that the nut will be below the surface of the wood.

For lightweight wood jambs, such as are generally used in houses, 20d nails can be driven into the jambs at each course and bent down over the shell as shown at *B*. The nails should be driven and bent before mortar is applied to the joints.

ANCHOR BOLTS

Houses and other buildings are no stronger than their weakest connections. Usually such connections occur where roofs are tied or anchored to walls and where frame walls are tied or anchored to foundations. Roof

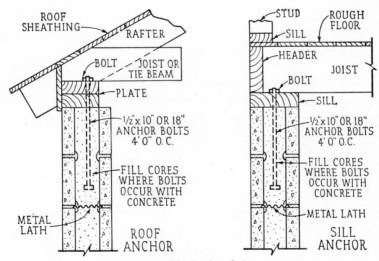

FIGURE 11. Anchor bolts.

connections are subject to severe strain, since winds tend to lift roofs upward as well as thrust them sideways. Therefore, it is important to provide rigid connections between them and walls. Frame sidewalls are also subject to strain because of wind pressure and to such unusual stresses as those produced by earthquakes. Therefore, it is also important to provide rigid connections between them and foundations. Figure 11 shows two typical examples.

Roof Anchors. EASTERN PRACTICE. When sidewalls up to roof levels are built of concrete block, one or two 2 by 6 or 2 by 8 wood plates are generally used to support the ends of joists or tie beams and rafters. Common practice is to attach the plates to the tops of concrete walls and to toenail the joists or tie beams and rafters to the plates. The wood plates should be anchored to the wall by means of ½-inch bolts which are at least 10 inches long and spaced not more than 8 feet apart. Some building codes require closer spacing. The cores of the two top courses of block in which the bolts are set must be completely filled with concrete to ensure good anchorage.

WESTERN PRACTICE. In regions where earthquakes are apt to occur, the building codes specify the use of *bond beams* at the tops of walls (bond beams are explained in succeeding pages). Such bond beams may be 8 or 16 inches deep and may have two or four steel rods in them. Anchor bolts are embedded in the bond beams, and the wood plates rest on top of the beams.

Sill Anchors. EASTERN PRACTICE. When frame sidewalls are supported by concrete-block foundations in buildings with basements, the sills (sometimes known as *mud* sills) are generally made of 2 by 6 or 2 by 8 timbers. Common practice is to secure the sills to the foundations by means of ½-inch bolts which are embedded at least 10 inches in concrete and spaced not more than 8 feet apart. Some building codes require closer spacing. The cores of the two top courses in which the bolts are set must be filled with concrete, as for roof anchors.

When setting anchor bolts, it is a good practice to use some sort of form, such as shown in Figure 12 of Chapter 6, Concrete Foundations. The bolts must be perfectly vertical and held in that position until the concrete hardens.

Sometimes solid or solid-top blocks, as shown at *L* and *H* in Figure 1 of Chapter 8, Concrete Block, are laid on the tops of foundations to support the floors and joists better. The solid-top block are 7⅝ inches in over-all height, and the top 4 inches are solid concrete. The solid blocks are generally 4 inches thick, as shown in Plate VI.

WESTERN PRACTICE. In regions where earthquakes are apt to occur, building codes specify the use of bond beams at the tops of foundations.

Such beams may be 4, 8, or 16 inches thick and contain two to four steel rods. Anchor bolts are embedded in, and the wood sills rest on, the beam.

PILASTERS

Like *concrete* pilasters, as explained in Chapter 6, Concrete Foundations, concrete-block pilasters generally have two purposes. First, they add strength to block walls or foundations at points where loaded beam ends

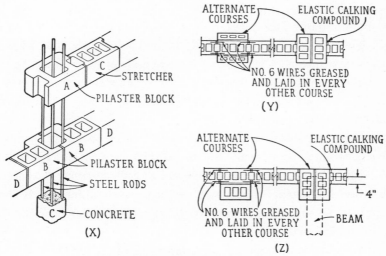

FIGURE 12. Typical pilaster construction.

must be supported as explained in Chapter 5, Concrete Footings. Second, they add stability to long walls and foundations. Figure 12 shows three typical ways in which pilasters can be laid up as parts of walls or foundations:

FIGURE 12 (*X*). Where extremely heavy beam loads must be supported, this type of pilaster construction works out to excellent advantage. The pilaster block, as shown at *A* and *B*, may be in one or two pieces and is used in connection with the kinds of ordinary stretchers shown at *C* and *D*. The *A* and *B* block should be used in alternate courses. Steel rods and concrete can be placed in the cores to the full height of the pilaster. This provides practically the same strength as though the pilaster was made of reinforced concrete.

FIGURE 12 (Y). Where ordinary beam loads, such as explained in connection with column footings in Chapter 5, Concrete Footings, have to be supported and for the purpose of wall or foundation stability, this type of pilaster construction serves a good purpose. Note that control joints are provided. The greased-wire reinforcing and the caulking compound allow some movement without damage to stiffness or alignment.

FIGURE 12 (Z). This type of pilaster construction serves the same purposes as the pilaster shown at Y.

Design of Pilasters. For all practical purposes in ordinary houses and other small buildings, the pilaster construction shown at Y and Z in Figure 12 is amply strong. For example, Plate II shows that an I-beam is necessary as a means of supporting interior ends of floor joists. The two ends of the beam must be supported by the block foundations. The required pilasters are 16 inches square. Thus, the construction shown at Y and Z would be satisfactory.

Where concrete-block walls and foundations of ordinary heights are involved, stability requirements can be satisfied (subject to local building codes) if pilasters, such as shown at Y and Z, are constructed at intervals of 18 times the nominal thickness of the walls or foundations. It should be kept in mind that cross walls serve the same purpose as pilasters where stability is concerned.

Most city and town building codes have definite regulations regarding the use of pilasters under all conditions. Readers are urged to check such requirements before planning any work of their own.

Where the loads of ordinary houses and other small buildings are concerned, simple pilasters can be checked for strength by using the formula

$$A = \frac{P}{f}$$

where A = required area of pilaster, square inches

$\quad P$ = load

$\quad f$ = allowable compressive working stress, pounds per square inch

Suppose, for example, that the house shown in Figures 9 and 10 of Chapter 5, Concrete Footings, has a concrete-block foundation composed of 8 by 8 by 16-inch stretchers and that one end of the beam between

X and Z has to be supported by such a foundation. Further suppose that the beam end is supported by the foundation without a pilaster. For purposes of illustration, it can be assumed that one lineal foot of the foundation must support the load from the beam end.

The area of 1 lineal foot of the foundation is 8×12, or 96 square inches. The load from one end of the beam is 17,100 pounds. The allowable compressive stress, as stated in Chapter 8, Concrete Block, is 70 pounds per square inch.

Substituting actual figures in the formula yields

$$A = \frac{17,100}{70} = 244$$

Thus, the foundation must supply a supporting area for the beam end of at least 244 square inches, whereas the actual area provided is 96 square inches. A pilaster is therefore necessary.

Next suppose that a pilaster such as shown at Z in Figure 12 is constructed under the beam and as part of the foundation. Then, the pilaster, being 16 inches square, provides an area for supporting the beam end of 256 square inches. The 256 figure is larger than $P = 17,100$ divided by $f = 70$. So the 16- by 16-inch pilaster is amply strong to support the load.

When large houses or other buildings are involved, a structural engineer should be consulted about pilaster design.

COLUMNS

Sometimes 8- by 8- by 16-inch concrete-block columns are used instead of wood or Lally columns. Where *light* loads are concerned and where the space required for a concrete-block column is of no concern, its use often brings about structural economy.

To all intent and purposes, a concrete-block column is the same as a concrete-block pilaster. As an example of column design, Figures 9 and 10 of Chapter 5, Concrete Footings, can be used. Column X supports the loads from two beam ends. The total load amounts to 34,200 pounds. Using the same formula and assuming a column constructed something like the pilaster shown at Z in Figure 12 (without the greased wire), the calculations are

$$A = \frac{P}{f}$$
$$A = \frac{34,200}{70} = 488$$

Thus, the column area in cross section must be at least 488 square inches. The 16-inch-square concrete block used for the pilaster provides only 256 square inches, which means that such a column is *not* strong enough. A larger column could be made by the use of three or more blocks per course, but such a column would require too much space and would have to be capped with several inches of placed concrete in order to distribute the load from the beams properly. Also, by filling the cores with concrete and using some steel rods, additional strength could be obtained. However, if an ordinary column such as was first considered cannot support the load, one of the two following alternatives must be employed:

First, consult a structural engineer for the design of a reinforced-concrete-block or reinforced-concrete column. Second, and by far the simplest solution, use a steel Lally column. Such columns can be selected directly from the manufacturer's published data. For example, a 3½-inch Lally column whose height (unbraced length) is not more than 8 feet can safely support a load of 32,300 pounds, and a 4-inch Lally column can safely support a load of 43,100 pounds.

WINDOW OPENINGS

When window openings are necessary in concrete-block walls and foundations, they must first be planned as described in Figure 6 and Tables 2 and 3 in Chapter 8, Concrete Block. Two kinds of block are commonly used for the vertical sides (jambs) of such openings.

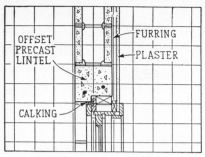

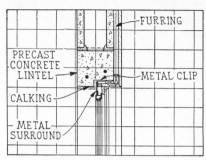

HEAD SECTION

HEAD SECTION

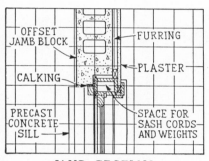

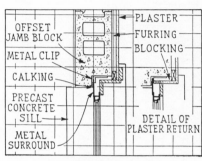

JAMB SECTION

JAMB SECTION

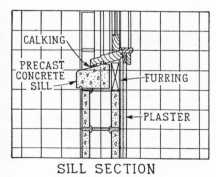

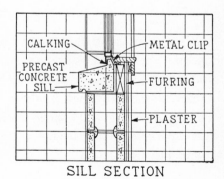

SILL SECTION

SILL SECTION

(A) DOUBLE HUNG WOOD WINDOW

(B) METAL CASEMENT WINDOW

FIGURE 13. Head, jamb, and sill sections for wood- and metal-sash windows.

Offset Jamb Block. A typical offset jamb block is shown at E in Figure 1 of Chapter 8, Concrete Block. Such blocks can also be secured in half lengths and with the end opposite the offset either a half core or square. The *A* and *B* parts of Figure 13 show how such blocks are used with wood double hung window frames (where sash cords and weights are required) and with metal casement windows.

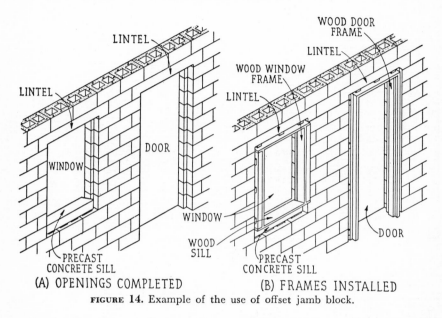

(A) OPENINGS COMPLETED   (B) FRAMES INSTALLED

FIGURE 14. Example of the use of offset jamb block.

When offset jamb blocks are to be used for the sides of window openings, the openings of the proper size are laid up in walls or foundations and the frames are inserted at some later time after all rough construction work is finished, and there is no danger of damaging the expensive frames. The *A* part of Figure 14 shows a window opening which was laid up before the frames were placed. The *B* part of the same illustration shows the frames installed.

Square Jamb Block. This kind of jamb block is similar to the block shown at *B* and *C* in Figure 1 of Chapter 8, Concrete Block. Such blocks can be secured in half lengths and with one end a half core or both ends square. The *B* and *C* parts of Figure 15 show how such blocks are used with wood double hung windows (which do not require cords and

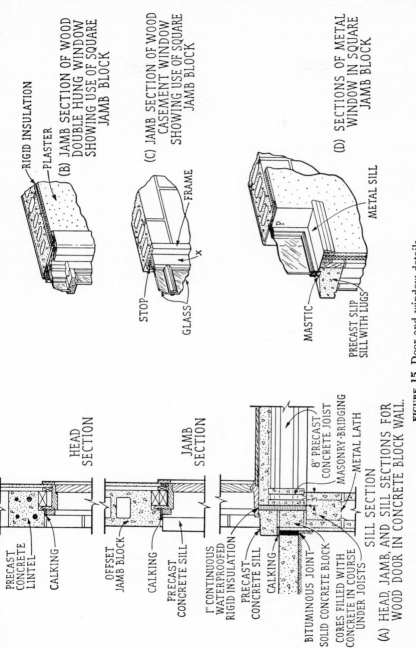

RIGID INSULATION
PLASTER
(B) JAMB SECTION OF WOOD DOUBLE HUNG WINDOW SHOWING USE OF SQUARE JAMB BLOCK

(C) JAMB SECTION OF WOOD CASEMENT WINDOW SHOWING USE OF SQUARE JAMB BLOCK
FRAME
STOP
GLASS
x

(D) SECTIONS OF METAL WINDOW IN SQUARE JAMB BLOCK
METAL SILL
MASTIC
PRECAST SLIP SILL WITH LUGS

PRECAST CONCRETE LINTEL
CALKING
HEAD SECTION

OFFSET JAMB BLOCK
CALKING
PRECAST CONCRETE SILL
JAMB SECTION

1" CONTINUOUS WATERPROOFED RIGID INSULATION
PRECAST CONCRETE SILL
CALKING
BITUMINOUS JOINT
SOLID CONCRETE BLOCK
CORES FILLED WITH CONCRETE IN COURSE UNDER JOISTS
8" PRECAST CONCRETE JOIST
MASONRY-BRIDGING
METAL LATH
SILL SECTION

(A) HEAD, JAMB, AND SILL SECTIONS FOR WOOD DOOR IN CONCRETE BLOCK WALL.

FIGURE 15. Door and window details.

weights) and with wood casement windows. There are various other types of wood frames which can be used with this kind of jamb block.

When square jamb blocks are to be used for the sides of window openings, the frames, as shown at x in the C part of Figure 15, are generally set in place before the openings are laid up. Figure 6 shows a window frame being placed in proper position prior to the time the opening is laid up.

With various types of spring balances available for use with double hung windows instead of sash cords and weights, use of square jam blocks with wood jambs fastened to the masonry jamb blocks as shown in Figure 10 is convenient. For inexperienced masons, square jamb blocks are easier to plan and to lay.

Jamb blocks having a narrow slit about 1 inch wide and in a vertical position are also available for use with metal sash. Such blocks are square, like those shown in the D part of Figure 15, and have slits as shown at P.

Blocks to fit the needs of any type of window can generally be secured. However, it is best to find what kinds are available before selecting window types.

When planning window openings, it is best to find what sizes of windows, of the desired kinds, are available, remembering that the window frame and concrete sill, of whatever kind used, must fit within horizontal and vertical dimensions of an opening which can be laid up without having to cut a whole or half block.

Door Openings. The kinds of jamb blocks used for window openings can also be used for door openings. When offset blocks are to be used, as shown in the A part of Figure 15, the openings of the proper size are generally laid up and the frames inserted at a later time. The A part of Figure 14 shows a door opening which was laid up before the frames were placed. The B part of the same illustration shows the frames installed.

When square jamb blocks are to be used for door openings, the frames, as shown in Figure 10, are generally placed before the openings are laid up. For inexperienced masons, the door opening and type of wood jamb shown in Figure 10 are the easiest to lay up and install.

BOND BEAMS

Builders are frequently confronted with local problems and requirements which need special construction procedures. For example, in earthquake regions it is necessary to provide for more than ordinary stability in concrete walls and foundations. This is also true in regions where heavy winds and severe storms are apt to occur. In fact, in almost every locality there are walls and foundations which are subject to one kind or another of unusual stress. Any unusual stress conditions require special attention by builders. Local building codes specify exact requirements which must be followed.

When bond beams are used at the tops of foundations, at the tops of one-story walls, and at each story height in higher walls, they add worthwhile horizontal reinforcement which helps materially to overcome stresses which might otherwise be dangerous. In principle, bond beams have tensile strength (because of the steel in them), as well as compressive strength. They help to support roof and floor loads better and to prevent walls and foundations from cracking or tearing apart. They tie walls and foundations together and add much greater strength and stability.

Kind of Block Used. The bond beam shown in Figure 11 of Chapter 8, Concrete Block, employs the kind of block shown at *K* in Figure 1 of that same chapter. Channel blocks, also shown in the same illustration, are sometimes used. Figure 16 shows a kind of block which was especially designed for bond beams. Note the knockouts. The block shown at *A* has three knockouts which can be removed by hammer blows to make the block look like the one shown at *B*. When necessary, the knockouts can be used to plug cores. The block shown at *C* has side knockouts so that it can be used at corners, as shown at *D*, where block *C* is used in the *y* position and block *B* is used in the *x* position. To form an 8-inch-deep block beam, the blocks are laid as shown at *D* and for a 16-inch-deep beam, as shown at *E*.

Channel blocks can be used for 8- and 16-inch beams in much the same manner as shown in Figure 16. However, two courses of such blocks are required for each 8-inch depth of beam.

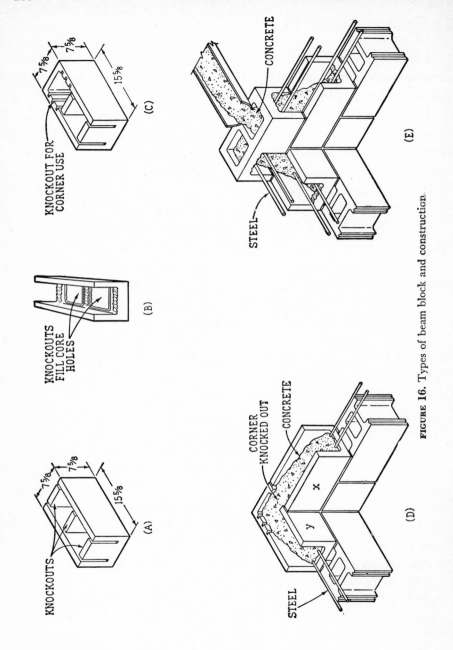

FIGURE 16. Types of beam block and construction.

General Requirements. Building codes vary in their specifications as to the kind of block to use, the amount of steel to place in the beams, and the depth of beams which must be used. Therefore, the following suggestions are for guidance only. Readers are urged to check actual requirements in their localities.

Foundation Block Beams. When foundations are used for basements, the tops of the foundations may have a bond beam either 8 or 16 inches in depth, as shown at *D* and *E* in Figure 16. Two ½-inch steel bars are used for each 8 inches of depth.

Wall Bond Beams. At each story height, walls may have a bond beam either 8 or 16 inches in depth as shown at *D* and *E* in Figure 16. Two ½-inch steel bars may be used for each 8 inches of beam depth.

JOIST SUPPORTS

In two-story houses and in other buildings where joist ends are supported by concrete-block walls, special blocks should be used in, around, and above the bearing area.

Wood Joists. The *A* part of Figure 17 shows a section view of a typical 8-inch block wall which supports the ends of ordinary wood floor joists. The slanting dashed lines at the end of the joist show how it should be cut so that the top edge cannot exert pressure on the wall if, for some reason, the joist should tilt. Note that the joist should have at least 4 inches of bearing on the wall.

When such a wall has been laid up to the underlevel of the joists, the cores in the block under the joists should be filled with concrete, using metal lath to plug the lower cores. The joists should next be placed in proper positions. Note that hangers should be nailed or bolted to the joists and bedded into the masonry work. Solid blocks having dimensions of 3⅝ by 15⅝ inches should be used for the outer facing of the joist-bearing course. The course above the joists should be laid using header blocks which have an offset over 2 inches deep. This offset allows the use of

10-inch joists. The space between the joists on the inner facing of the wall, known as *bridging,* should be filled in, using partition blocks which can be 3⅝ inches thick. Above the joists, the bridging can be filled using concrete brick or frogged brick of the kind shown in Figure 1, of Chapter 8, Concrete Block. If joists no higher than 7⅝ inches (most 2 by 8 joists actually measure only 7⅝ inches) are to be used, a somewhat simpler method of construction can be employed. After the cores have been filled

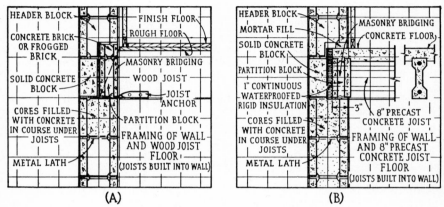

(A)   (B)

FIGURE 17. Joist supports in concrete-block wall.

in the course upon which the joists are to rest, the joists should be set in their required positions. Then jamb blocks can be placed around them. The offset in the jamb block allows ample room for the joist bearing. Above the jamb blocks, the next course can be laid, using regular block.

Precast Concrete Joists. Joists of this type are supported by block walls in much the same manner as wood joists, as shown by the *B* part of Figure 17. The only difference is that precast joists need but 3 inches of bearing and rigid insulation back of the outer face and that concrete brick is not needed. The concrete floor should be placed before the header course is laid up.

Where either wood or precast joists run parallel to walls, one joist should be placed as near the wall as possible so that the floors will be supported at the points where they meet the walls.

METAL BASEMENT WINDOWS

In many instances where concrete-block foundations are employed, such as shown in Plates II and VI, the basement windows have metal frames and sash. A typical window of this kind is shown in Figure 18. Note that in the case of basement windows a wider precast sill is used

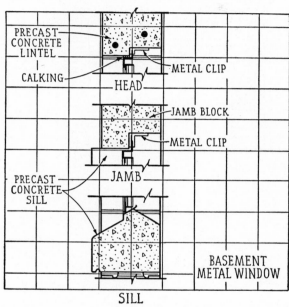

FIGURE 18. Head, jamb, and sill sections of metal basement window.

than is shown in Figure 7. Note, too, that the lintel and jamb blocks may have offsets or fins in them.

INTERSECTIONS

Care should be taken to bond or tie all intersections securely where walls meet walls or foundations meet foundations. The following requirements are typical:

EASTERN PRACTICE. When two *bearing* walls or foundations (load supporting) meet or intersect and the courses are laid up at the same time, a true masonry bond between at least 50 per cent of the units at the intersection is necessary. This is shown in Figure 19.

When such intersecting walls or foundations are laid up separately, cavities with 8-inch maximum vertical offsets are left in the first wall

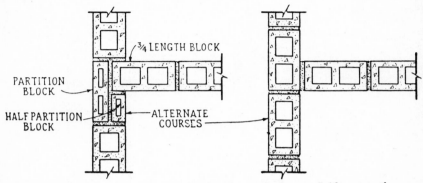

FIGURE 19. Plan view of intersecting bearing walls that are laid up together.

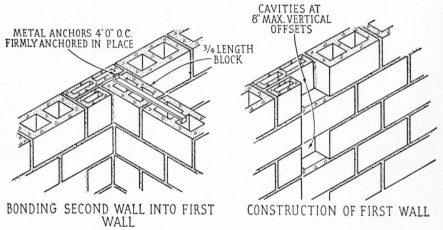

FIGURE 20. Bonding of intersecting bearing walls that are laid up separately.

or foundation laid up. The corresponding anchors, as shown in Figure 20, must securely engage the masonry to provide good anchorage.

WESTERN PRACTICE. In regions where earthquakes are apt to occur, most building codes specify that the courses of the intersecting walls or foundations shall be laid up at the same time and that two steel rods, in alternate courses, must be bent so as to be from 12 to 24 inches in both walls or foundations.

PARTITION AND BEARING-WALL OR FOUNDATION INTERSECTIONS

Ordinarily, partitions do not carry loads and are used merely as a means of dividing floor space into rooms.

EASTERN PRACTICE. Figure 21 shows how metal lath should be used in every second course as a means of tying partitions to walls or foundations.

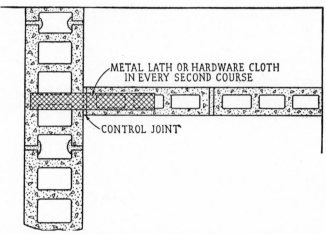

FIGURE 21. Method of bonding partition with a wall or foundation.

WESTERN PRACTICE. In regions where earthquakes are apt to occur, most building codes specify that at least one steel rod, in alternate courses, must be bent so as to be from 12 to 24 inches in the partitions and walls or foundations.

CONCRETE FLOORS SUPPORTED BY BLOCK WALLS

When a combination of concrete floors and precast joists is to be supported by block walls, the recommended construction is as shown in the *B* part of Figure 17. An alternative is the use of solid-top block in place of the cored block filled with concrete.

When concrete floors without precast joists are to be supported by block walls, the recommended construction is as shown in Figure 22. Note that insulation should be used between the solid-face block and the edge of the floors. The course at *x* can be laid up using cored block filled with concrete or by using solid-top block.

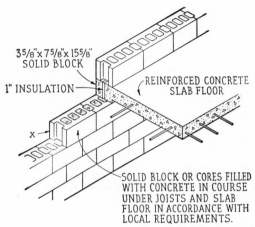

FIGURE 22. Concrete floor supported by block wall.

I-BEAMS SUPPORTED BY BLOCK WALLS

From the previous discussion about the strengths of concrete-block pilasters and columns, it can be seen that extreme care should be exercised in the design of pilasters to support such beam ends as shown in Plate VI.

In Plate VI, one of the two I-beams supports half the floor area between its pilaster and the chimney. The length of that area is 9′ 4″ plus 9′ 10″, or 19′ 2″. The width of that area is 11′ 10″. For ease in calculation, the length and width can be called 19′ 0″ and 12′ 0″. The area is 19 by 12, or 228 square feet. Half of that is 114 square feet. Assuming that the floor supports a total dead and live load of 70 pounds per square foot, the total load on the beam is 114 × 70, or 7,980 pounds. The pilaster must support half of that (the chimney structure supports the other half), or 3,990 pounds. From the previous discussion, it is evident that the block foundation, even without a pilaster, can safely support such a load. However, in order to be on the safe side, the pilaster was decided upon. Where concrete-block structures are concerned, it is always best to be on the safe side. Then, too, the pilaster in the foundation next to the unexcavated area will give the foundation lateral support in resisting any possible pressure from the soil back of it.

Figure 23 shows one method of construction around the ends of I-beams supported by block foundations. All cores for a distance of two

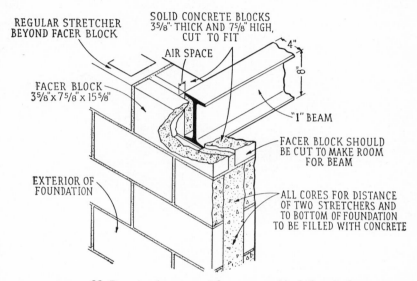

FIGURE 23. Beam end supported by concrete-block foundation.

stretchers under the beam and to the bottom of the foundation can be filled with concrete. The beam end should have a bearing of at least 4 inches. Once the beam has been set into position, a facer block can be laid in regular bond. Another facer block can be cut so as to fit around the beam on both sides, to complete the foundation thickness and to help keep the beam erect.

In cases where pilasters are required to support beam ends safely, the beams may be set on top of pilasters so that their bearings extend at least 4 inches into the walls, as shown by the dashed lines in the Z part of Figure 12.

Building codes and practices vary considerably in regard to the methods of bedding such beams on foundations and pilasters. Therefore, readers are urged to check codes in their localities.

DUCTS IN CONCRETE-BLOCK WALLS

Many of the sheet-metal ducts used in heating and air-conditioning systems for houses and other small buildings are 3⅝ inches thick. Thus,

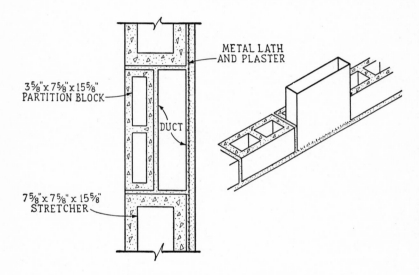

FIGURE 24. Duct in 8-inch wall with plaster over duct.

they can be built into block walls and foundations, as shown in Figure 24. The use of partition blocks maintains the bond and good appearances. Ducts thicker than 3⅝ inches and wider than 15⅝ inches should not be used in 8-inch block walls.

NAILS FOR USE WITH CONCRETE BLOCK

Nails that are tempered and hard enough to be driven into concrete block without damage to themselves or block are readily available. Their use avoids the necessity for inserting nailing strips into mortar joints and saves a great deal of time.

REINFORCEMENT

Concrete-block masonry is often more convenient and economical than cast-in-place concrete for the construction of walls, foundations, chimneys, retaining walls, and a host of other structural details. Such masonry material allows quick erection and can easily be made neat and attractive. For many purposes, it is an ideal building material, and its use is increasing by leaps and bounds.

When concrete-block masonry was first introduced, and until recent years, it had some limitations which prevented the wide usage now possible. Such limitations had to do with its lack of tensile strength and its tendency to crack or come apart at the joints when other than straight-down compressive stresses or loads were applied to it. The lack of tensile strength constituted the same drawback from which cast-in-place concrete suffered, until the use of steel reinforcement and prestressing solved the problems. The tendency to crack and pull apart at joints was a problem common to all kinds of unit masonry. Such limitations have been overcome by the proper use of steel reinforcement. Now, concrete-block walls and foundations can be given the necessary tensile strength and the ability to resist stresses which would otherwise cause trouble.

The use of reinforcement was probably developed to a great extent in sections of the country where earthquake stresses have to be carefully considered in all structural work. Yet, other sections of the country suffer windstorms and tornadoes which create equally important problems in structural work. As previously noted, there is scarcely a locality in which some form of unusual stressing is not likely to occur. Even the alternate freezing and thawing which occur in Northern localities bring about stressing which should not be overlooked. Therefore, it is wise to reinforce *all* concrete-block construction no matter in what locality or section of the country it may be.

City and town building codes contain strict rules pertaining to proper reinforcement. Most such rules are covered by the following suggestions.

But readers are urged to check requirements in their localities and to make absolutely sure that they make the most of the great benefits of proper reinforcement.

Footing Reinforcement. In most instances, as explained in Chapter 5, Concrete Footings, a footing 20 to 24 inches wide in average soil satisfies all compressive-loading requirements and can be counted upon to support foundations, columns, etc., satisfactorily. It is only when severe

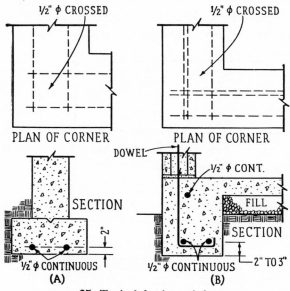

FIGURE 25. Typical footing reinforcement.

wind or earthquake stresses have to be considered or when weak spots in the soil are possible that plain concrete might fail. Thus, the use of two or more ⅜- or ½-inch steel rods, which are at least 2 inches above the bottom of a footing, which are parallel to its length, and which extend the full length of the footing, constitutes what can be called safe practice. Figure 25 shows typical examples.

Joint Reinforcement. The *A* part of Figure 26 shows a portion of a block wall or foundation which is reinforced in several of the horizontal joints; the *B* part of the illustration shows what the reinforcement looks like; and the *C* part, how it can be cut and bent at corners. Such re-

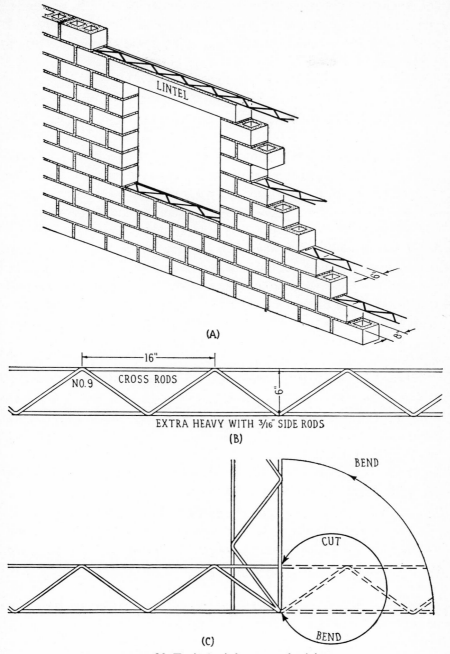

FIGURE 26. Typical reinforcement for joints.

inforcement should be used above and below openings in walls or foundations in addition to use in several horizontal joints.

The reinforcement can be embedded in joints and adds great strength to the structural details.

RETAINING-WALL REINFORCEMENT

The *A* part of Figure 27 shows the outline of a typical retaining wall whose function is to hold a soil bank in a desired position. It should be understood that such a soil bank, especially when wet, constitutes a great unstable force which tends to bow out or push the retaining wall.

Without ample and proper steel reinforcement, the wall is likely to fail in one or both of two ways. First, as shown at *D*, the pressure from the soil bank may bow out the wall and cause it to collapse. Second, the pressure may break the mortar joints near the footing and cause the wall to slide, as shown by the dashed outline at *B*.

In order to make such a wall strong enough to resist pressures of this sort, dowels and vertical rods (sometimes known as *studs*) should be built into it.

When the footing is placed, and before it has hardened, 24-inch dowels (generally ½- or ¾-inch steel rods) should be inserted into it, as shown at *G* in the section and elevation views. The dowels should be spaced not more than 24 inches apart. As the first two courses are laid up, the cores which contain dowels should be filled with pebble concrete. Where cores contain dowels, the faces of the block should be marked so that, when the wall has been laid up to its full height, the rods, as shown at *C*, can be inserted into cores adjacent to the dowels. The vertical rods, generally ½ to ¾ inch in diameter, should extend to the full height of the wall and be inserted at 24-inch intervals. The cores should be filled with the pebble concrete.

Garden Walls. Concrete-block garden walls should be reinforced in order to prevent cracks which could otherwise be caused by heaving or earthquakes.

The footings should contain two ⅜- or ½-inch steel rods placed as

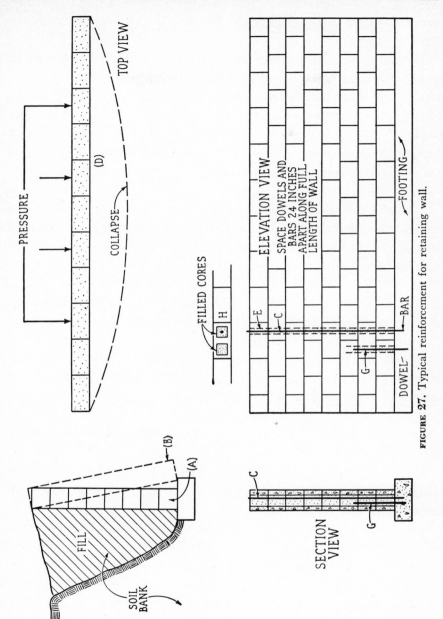

FIGURE 27. Typical reinforcement for retaining wall.

shown in the *A* part of Figure 25. The steel is not too costly and is a good investment.

The walls above the footing should contain ½-inch vertical steel rods, spaced not more than 4' 0" apart. They should be anchored in the footings in a manner similar to the dowels shown in the *B* part of Figure 25. Care should be taken to make sure that the rods, which are set before the blocks are laid up, will be at core locations in the walls. This layout can be accomplished by placing a row of blocks along the side of the footings so that core locations can be marked.

Wall and Foundation Reinforcement. Figure 11 of Chapter 8, Concrete Block, shows many typical aspects of reinforcement. Other aspects are discussed in foregoing pages of this chapter. Therefore, the following information constitutes a review plus a few added considerations.

Bond Beams. In general, two to four steel rods are required, depending upon whether such beams are 8 or 16 inches deep.

Dowels. Dowels, as shown in Figure 25, are generally spaced at 24-inch intervals. They should be embedded in the footings as shown and should extend up above the footings at least 24 inches.

Vertical Rods. There are several locations where vertical rods (studs) should be used.

Two rods, generally ½ inch in diameter, should be used at the corners. They should extend into the bond beams and lap the dowels at least 24 inches.

Two of them, generally ½-inch, should be used in the jamb-block cores on both sides of all window and door openings. They should extend into the bond beams and lap the dowels at least 24 inches.

One of them, generally ½-inch, should be placed at 24-inch intervals all the way around walls or foundations. They should extend into the bond beams and lap the dowels at least 24 inches.

Horizontal Rods. Two such rods, generally ½-inch, should be placed in the first mortar joint, one course below the sills, for all window openings.

The rods should be about 24 inches longer than the width of the openings so that they extend beyond the opening on both sides. For this purpose, the reinforcement shown in Figure 26 can also be used.

Intersections. As explained for Western practice in connection with Figure 20, all intersections of load-bearing walls and foundations must have at least two rods, generally ½-inch, which extend into the intersections of both walls or foundations. For this purpose, the reinforcement shown in Figure 26 can also be used. In addition, vertical rods can be used in adjacent cores, as shown at *B* in Figure 11 of Chapter 8, Concrete Block, and tied together with metal ties in every other course. All cores in which rods are placed must be filled with concrete.

CORNER CONSTRUCTION

In all concrete-block construction, the corners (leads) are laid up first. This is explained in Chapter 10, Concrete-block Walls, Foundations, and Pilasters. Several details are shown in Figure 28.

FIGURE 28 (*A*). This illustration shows how corners are laid up using ordinary sizes and shapes of blocks. The bond is easy to maintain by simply reversing every other course.

FIGURE 28 (*B*). For small retaining and ornamental garden walls, thin construction is often desirable. L-shaped blocks with rounded corners are available for such projects. The L-shaped blocks are simply reversed in alternate courses.

FIGURE 28 (*C*). When slightly heavier and thicker walls are desired, neat-appearing corners can be laid up using a 6-inch block, along with *L* corners.

FIGURE 28 (*D*). Where cavity walls are concerned, the corner is constructed by using a half-length block in alternate courses.

A simple method of preventing the accumulation of mortar droppings between walls and of maintaining a clear cavity is to lay a 1- by 2-inch board across a level of ties to catch the droppings. As the masonry reaches the next level for placing ties, the board can be raised, cleaned, and again laid on the ties.

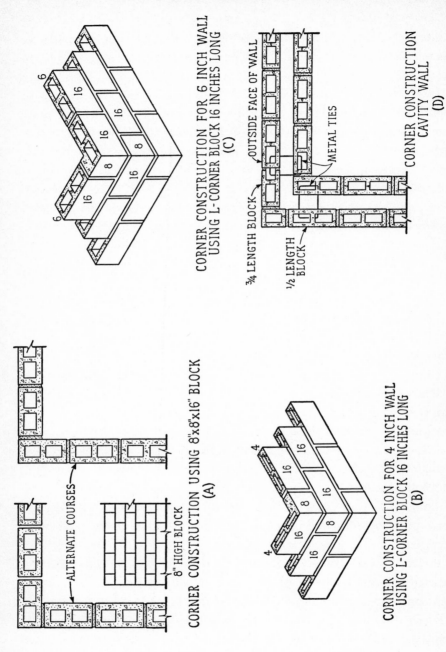

CORNER CONSTRUCTION USING 8"x8"x16" BLOCK
(A)

ALTERNATE COURSES

8" HIGH BLOCK

CORNER CONSTRUCTION FOR 6 INCH WALL
USING L-CORNER BLOCK 16 INCHES LONG
(C)

CORNER CONSTRUCTION FOR 4 INCH WALL
USING L-CORNER BLOCK 16 INCHES LONG
(B)

OUTSIDE FACE OF WALL

METAL TIES

¾ LENGTH BLOCK

½ LENGTH BLOCK

CORNER CONSTRUCTION
CAVITY WALL
(D)

FIGURE 28. Typical corner construction.

The practice of providing *weep holes* (passages through which water can escape) presumes that water will enter the wall from the outside. If the blocks are properly laid up and the mortar joints well compacted, the walls should be waterproof. However, in some localities subjected to severe driving rain, the use of weep holes is recommended.

The holes can be placed every two or three blocks apart and in the vertical joints of the bottom course on the outside wall immediately above the foundation or footing. In no case should the holes be located below grade. The holes can be formed by placing a well-oiled piece, or pieces, of rubber tubing in the mortar joint and then extracting it after the mortar hardens. The tubing should extend up into the cavity for several inches to provide a drainage channel through any mortar droppings that might have accumulated.

WATERTIGHT FOUNDATIONS

Concrete-block foundations can be made watertight if careful attention is given to making good mortar joints and if other structural precautions are heeded.

Joints. When mortar is applied to the shells of blocks, care should be taken to see that the shells are well covered to a thickness of at least 1 inch. Excess mortar can be recovered as blocks are placed. Careful tooling of joints also helps to make them watertight under normal circumstances. Concave and V-shaped joints are recommended. Struck and raked joints are apt to have small ledges on which water can gather. With modular-sized block, mortar joints will be about ⅜ inch thick. Experience has shown that this thickness, when properly made, helps to produce a neat, watertight, and durable foundation.

Exterior Surface. In localities where the subsoil is wet, where natural drainage is not good, or where heavy rains occur, the exterior faces of foundations should be given two ¼-inch-thick coats of plaster, using the same mortar as for the joints. In hot and dry weather, the faces of the foundation should be lightly sprayed with water prior to the application of the first coat. The first coat should be roughened after it has partly hardened to provide bond for the second coat. It should be allowed to harden for at least 24 hours before the second coat is applied. The first coat should be lightly dampened before the second coat is applied. The second coat should be frequently sprayed to keep it damp for 48 hours after application.

The surface of the plaster may be given two continuous coatings of hot bitumen. That material should be brushed or mopped on so as to form a solid film.

Figure 17 in Chapter 8, Concrete Block, and the explanations relative to it describe a method of using draintile around footings. Where extremely wet subsoil is known or suspected, such drainage is worth the additional cost.

WATERTIGHT WALLS

Watertight block walls are obtained by a combination of proper planning, the use of good materials, and good workmanship. When planning walls, proper flashing should be provided where water, snow, or ice may accumulate. Flashing is necessary at vertical points in coping and caps, at the joints between roof and walls, and below cornices and other members which project beyond the face of walls. In addition, drips should be provided for chimney caps, sills, and other projecting ledges to shed water away from the walls. Drains and gutters must be large enough to keep water from overflowing and running down over wall areas.

Good workmanship is always an important factor in building weathertight walls. Each block should be laid up plumb and true, with care that both horizontal and vertical joints are well filled and compacted by tooling after the mortar has partly stiffened.

Concrete-block walls can be made more watertight by the use of any one of the numerous paints and compounds now available.

COPING

All exposed tops of walls, garden walls, retaining walls, etc., should have a coping block for the full length. Such blocks can be obtained in various shapes and colors. Generally they are applied to the tops of walls with mortar.

WALLS AROUND COLUMNS

It often happens that block walls have to be laid up around columns such as shown in Figure 29. This illustration is self-explanatory.

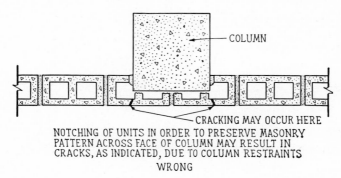

NOTCHING OF UNITS IN ORDER TO PRESERVE MASONRY
PATTERN ACROSS FACE OF COLUMN MAY RESULT IN
CRACKS, AS INDICATED, DUE TO COLUMN RESTRAINTS
WRONG

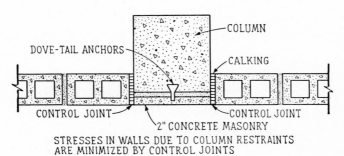

STRESSES IN WALLS DUE TO COLUMN RESTRAINTS
ARE MINIMIZED BY CONTROL JOINTS
RIGHT

FIGURE 29. Wrong and right methods of constructing concrete-block walls around columns.

CHAPTER TEN

Concrete-block Walls, Foundations, and Pilasters

In Chapter 8, Concrete Block, and Chapter 9, Typical Concrete-block Details, all the commonly encountered fundamentals and practices pertaining to ordinary concrete-block planning and construction were explained and illustrated. Those two chapters should be carefully studied and thoroughly understood before a study of this chapter is undertaken.

The purpose of this chapter is to show how such fundamentals and details are actually used to plan and lay up walls, foundations, and pilasters. The application of the fundamentals and recommended details, while not difficult, is somewhat intricate, especially for inexperienced masons, and should therefore be given strict attention. Most of the illustrations in this chapter are of the type which demonstrate various aspects of the work, as well as illustrate principles. The methods shown can be applied to practically any concrete-block project which readers are likely to encounter.

PLANNING AND ERECTION CARE

As previously pointed out, concrete-block construction is very economical, durable, and completely satisfactory for many kinds of houses and other buildings located in almost any climate or region. However, because this type of construction involves the use of comparatively small units, and because of the large number of joints between such units, the

208

greatest possible care must be exercised in both the planning and erection stages of any project.

Planning. No concrete-block project of any appreciable size or scope can successfully be carried out unless or until accurate planning has been done using a scale, such as $\frac{3}{4}'' = 1'\ 0''$, and some means of drawing sketches such as those shown in Figures 2, 3, 4, 5, and 19. The sketches need not be works of art, but they must be accurate so that each block of the structure being planned will fit into the proper place without the necessity for cutting block, ruining bond, or otherwise causing poor or unsafe construction. Mental pictures should not be trusted because of the chance of omitting or overlooking important aspects of the planning. Only careful drawings and calculations assure good results. In fact, city and town building departments generally refuse to issue permits unless projects are shown in ample detail by drawings. The great value of drawings is shown in Chapter 14, Illustrative Example. Such drawings are not difficult for masons to make, and they assure the success of any project.

Workmanship. The necessity for good workmanship has also been previously mentioned. The best possible planning will be to no avail unless the planned work is carried out according to the rules of good craftsmanship. This means taking pains to make sure that each block is laid up accurately; that all joints are full, well-tooled, and packed; that all work is plumb in all directions; that reinforcing steel is properly set; and that, where necessary, the cores are completely filled with good concrete. Poor work seldom passes building-department inspection. Even worse, poor work never fails to cause dissatisfaction and loss of money. Inexperienced masons should proceed slowly and cautiously until experience has been gained.

PLAN OF CHAPTER

In order to illustrate and explain how walls, foundations, and pilasters are planned and laid up, the plans and details shown in Plates II and VI will be used. Actually, these plates have to do with *foundations*, along with two pilasters. However, the procedures for both the planning and erection of such foundations apply equally well to walls above grade. In many instances, walls are supported by footings in the same manner as the foundations shown in the plates. Also, the pilasters built into walls are the same as those built into foundations. In other words, the following sections of this chapter can be thought of in terms of both foundations and walls and the pilasters built into them. This book makes a distinction between walls and foundations only because most readers probably think of the two structural details in a somewhat different light. But so far as concrete-block construction is concerned, both details are built in the same manner.

Two typical examples of concrete-block planning and construction are presented in this chapter. One example has to do with *nonreinforced* construction and the other with the *reinforced* construction which is required in many sections of the country. The same details are used for both examples in order to illustrate and explain better the typical procedures and to avoid the necessity for readers having to visualize more than one general situation. Because of the growing importance of reinforcement in concrete-block construction, readers are urged to give the subject careful attention, even though their local building codes have not yet been changed to include steel in such construction.

Certain aspects of both examples are assumed. For example, the design and construction of footings are not explained because such procedures are described in Chapter 5, Concrete Footings. The mixing of mortar is not explained, because such material is included in Chapter 3, Mortar. Two-core square-end blocks of the 8- by 8- by 16-inch nominal size are assumed because they are somewhat easier to illustrate and handle in actual work. All the planning aspects follow typical building-code specifications, and readers will find that they can interpret both of the examples in terms of their local code requirements.

The foundation design shown in Plate II is not completely modular because the architect meant the plans to be used with or without modular materials. Both examples include such dimension changes as are necessary to come within modular procedure. It is best to learn the modular procedures because more and more materials, especially masonry materials, are being produced in such sizes and shapes. The modular procedures do not in any way detract from general education, but they do make great economy and improvement in structural work possible. Once a reader has learned the modular procedures, he can easily adapt them to any size and shape of concrete block.

EXAMPLE 1   (No reinforcement)

Figure 1 shows part of the foundation plan of Plate II. The *ABCDEG* area includes the eastern half of the general basement plan. This half is to have a full basement. The other half is to have a crawl space under the floor. The height of the foundation in this area, as shown by the *W*-1 detail in Plate VI, is to be 7′ 8″ above the footings.

Planning. Plate II indicates that the distance between *A* and *B*, as shown in Figure 1, is 5′ 8″ + 7′ 6″, or 13′ 2″. This is not a modular dimension because, according to Table 1 in Chapter 8, Concrete Block, the nearest modular dimension is 13′ 4″. In other words, a 13′ 4″ dimension allows *exactly* ten stretchers when 15⅝-inch modular blocks are used. Therefore, for the purpose of this example and to follow good planning procedure, the distance *AB* has been changed to exactly 13′ 4″.

Plate II indicates that the distance from *B* to *C*, as shown in Figure 1, is 5′ 6″ plus the width (8 inches) of the foundation, or 6′ 2″. This is not a modular dimension because according to Table 1 the nearest modular dimension is exactly 6′ 0″. Therefore, the 6′ 0″ dimension will be assumed, as shown in Figure 1.

Plate II indicates that the distance from *C* to *D* in Figure 1 is 7′ 4″ plus the width of the foundation, or exactly 8′ 0″. Table 1 shows that this is a modular dimension. It will therefore be used as shown in Figure 1.

The foregoing aspect of planning shows the necessity for starting with proper dimensions when drawing the plans for houses or other buildings. In most cases, a matter of a few inches makes the difference between modular (economical) construction and nonmodular (uneconomical) construction. This discussion also shows that masons should study whatever plans they encounter to see if the dimensions are modular. Masons are not responsible for architect-prepared plans, but sometimes just a little cooperation helps any building project.

Plate II indicates that the pilaster center line is 2′ 3″ from the north section of the foundation. Adding this dimension to the thickness of the foundation gives 2′ 11″, which is not a modular dimension. The pilaster, as shown by the *Z* part of Figure 12 in Chapter 9, Typical Concrete-

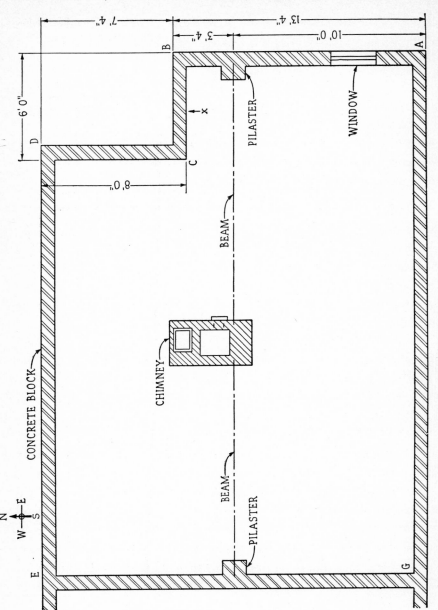

FIGURE 1. Partial plan of Plate II.

block Details, will be composed of two whole blocks. Therefore, the distance from corner *B,* as shown in Figure 1, must be a modular dimension. The nearest such dimension is 3′ 4″ or 2′ 8″. The 3′ 4″ dimension is assumed.

In actual practice, such dimension changes would have to be made by architects, so that other details of the building would coincide. However, when masons are planning their own buildings, as often happens, especially in rural areas, dimension changes should be made only with careful thought to other parts of buildings. Better yet, make the dimensions modular in the first place, as explained in Chapter 8, Concrete Block.

With the modular dimensions shown in Figure 1 as a basis, further planning can be done which will assure good construction procedures. The following illustrations show how masons can plan their work in such a way that they will know exactly what to do so far as actual erection is concerned.

Figure 2 shows several typical courses of the foundation on side *AB.* This kind of planning should be carried out by all inexperienced masons as a means of helping them to plan accurate work and to give them a sure method of visualization. The courses in Figure 2 are drawn to scale. Thus, it can be seen that ten of the 16-inch stretchers (15⅝ plus ⅜ joint) equal *exactly* 13′ 4″. Note, too, that the joint between pilaster blocks is exactly 3′ 4″ from corner *B* and that this dimension includes whole stretchers or half stretchers as indicated for the second course. After the first course, all courses are alternated, as shown by the first and second courses.

Plate II indicates that a window is necessary in the *AB* wall shown in Figure 1. The 5′ 8″ dimension shown in the plate is not modular. Therefore, as shown in the eighth course in Figure 2, the left jamb of the window is assumed as being 32 inches (two stretchers) from the corner *A* shown in Figure 1. Note, too, that alternate courses may have half blocks in them. The width of the window opening is assumed to be equal to two stretchers and the height, according to Table 2 in Chapter 8, Concrete Block, equal to 2′ 8″, or exactly four courses.

Figure 3 shows the course plan for the side *AB* of Figure 1. This kind of planning, used with block plans as shown in Figure 2, is also of great help to inexperienced masons.

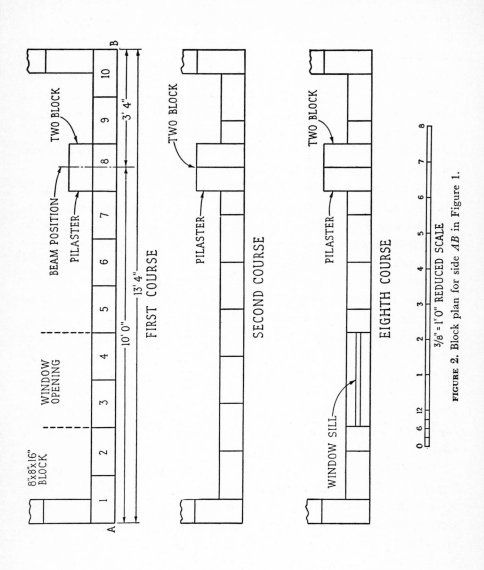

FIRST COURSE

SECOND COURSE

EIGHTH COURSE

3/8" = 1'0" REDUCED SCALE

FIGURE 2. Block plan for side *AB* in Figure 1.

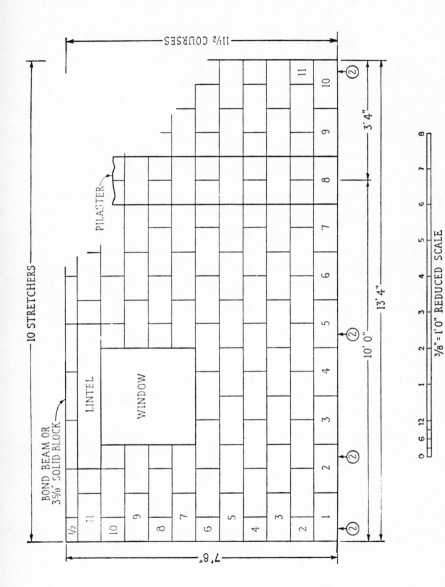

FIGURE 3. Course plan for side *AB* in Figure 1.

The W-1 detail in Plate II indicates that the height of the foundation should be 7′ 8″ above the footings. This height is equal to exactly 11½ courses and works out as a modular dimension because, according to the W-1 detail, the top course should be composed of solid half block. By drawing the course plan to scale, as shown in Figure 3, the AB side of the foundation can be planned and checked. The scaled drawing proves that the pilaster and window-opening plans made in connection with the block planning (Figure 2) are correct. The course plan also shows that the lintel over the window opening has ample and proper bearing and that whole block can be used and proper bonding assured.

Figure 4 shows the block plans for sides CB and CD, as shown in Figure 1. In order to visualize the block plan for side CB, it is necessary to imagine that the side is being viewed from the position of the x arrow shown in Figure 1. The half block shown at corner B in Figure 4 is the same as block 10 shown in the first course illustrated in Figures 2 and 3. With four more whole blocks the distance of 6′ 0″ (4½ stretchers) is exactly made up. The second course for side AB is the opposite of the first course. The block shown at corner B is the same as the block shown at 11 in Figure 3.

Figure 5 shows the course plan for side BC in Figures 1 and 4. This plan checks out the block plan with the required height of the foundation. The half blocks (ends) which are shaded are those blocks, such as block 10, in the first course of side AB, which are full-length stretchers at the right-hand end of side AB.

The block and course plans shown for side DC in Figures 4 and 5 also check out these plans. The half blocks (ends) which are shaded in Figure 5 are those blocks which are full stretchers in the side BC.

Readers are urged to study Figures 2 through 5 until they can easily visualize the block and course plans in terms of the foundation sections (sides) mentioned. Once such visualization has been mastered, the planning of any concrete-block project is easy.

The balance of the foundation sections shown in Plate II can be planned in the same manner as explained for sides AB, BC, and CD in Figures 2 through 5. With such plans before him, an inexperienced mason will have no trouble so far as layout of blocks is concerned.

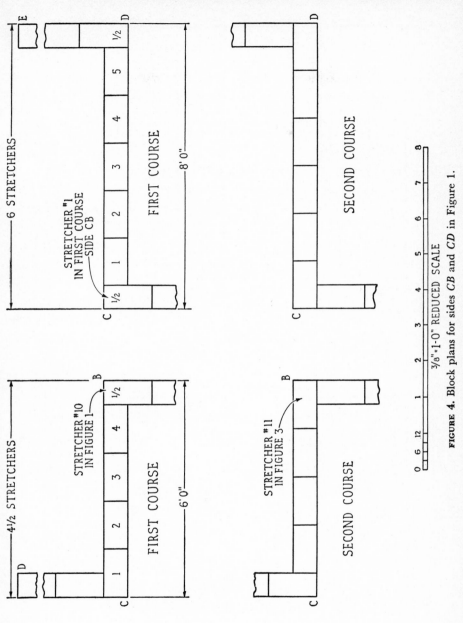

FIGURE 4. Block plans for sides *CB* and *CD* in Figure 1.

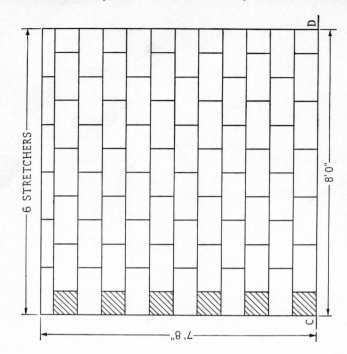

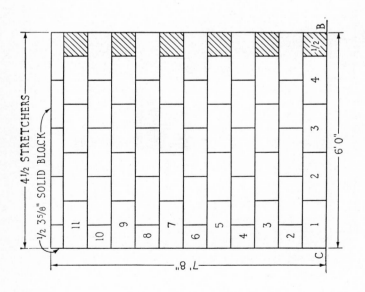

FIGURE 5. Course plans for sides CB and CD in Figure 1.

Mortar Application. There are three places where mortar is applied as blocks are being laid up. It is important for inexperienced masons to learn the proper procedure.

MORTAR ON FOOTINGS. When mortar is applied directly to the surface of footings, as a *bed* for the first course of block, it should be spread so as to cover completely the area the block is to cover. Figure 6 shows a mason applying mortar to a footing prior to laying a block in a first course. Note that the mortar makes a full bed and that it is slightly furrowed with the

FIGURE **6.** A full mortar bed should be placed on footing. (Courtesy of Portland Cement Association.)

trowel. The bed should be at least 1½ inches thick even though some of the mortar is squeezed out as the blocks are laid. There should be no skimping of mortar, and the bed should be under and longer than the block.

MORTAR ON BLOCK ENDS. Most experienced masons prefer to apply mortar (known as *buttering*) to the ends of the block as shown in Figure 7. The block should be stood on one end and the mortar (sometimes called *mud*) applied as shown. There is no necessity or reason for covering the entire end of the block. This mortar should be about 2 inches thick and spread a little after being applied to the block. Some of the mortar will be squeezed out as the block is pressed into position.

FIGURE 7. Applying mortar (buttering) to end of a block. (Courtesy of Portland Cement Association.)

FIGURE 8. Bed joints laid with face-shell mortar. (Courtesy of Portland Cement Association.)

BED JOINTS ABOVE FOOTINGS. After the first course of block has been laid, the mortar for bed joints under all succeeding courses is applied as shown in Figure 8. Note that the mortar, about 1½ inches thick, is applied only to the face shells in two parallel rows. The illustration also shows the end of a block with mortar in place. Some masons like to apply such vertical-joint mortar to blocks already in position instead of as shown in Figure 7. Either procedure constitutes good practice.

The application of mortar requires considerable practice, and inexperienced masons should not be discouraged if at first they drop or waste more mortar than they actually apply.

The mortar for bed joints should not be spread for more than one block at a time because it tends to stiffen quickly and thereby loses some of its plasticity. When this happens, poor joints are often the result, the joints lose strength, and walls or foundations thus become weaker and less likely to resist the penetration of moisture, wind, and dirt.

Leads. It is the usual practice and the best policy to lay up block walls or foundations by first laying up the corners, three or four courses high, and using them as guides for intervening blocks. In other words, for every three or four courses, all the way around the walls or foundations, the corners (known as *leads*) are first laid up. This procedure is explained in subsequent pages and illustrated in Figures 9, 10, 12, 13, and 15.

Preparation. The *W*-1 detail in Plate VI indicates that the foundation shall be placed on the 20-inch-wide footing so that the footing extends 6 inches beyond the foundation on both sides. Thus, the first preparatory step is to draw pencil or chalk guide lines on the footing, as shown in the *X* part of Figure 9. Such guide lines must be accurately drawn and should be checked most carefully. They help materially in laying the first course accurately on the footings. The lines should be located with an allowance of $\frac{3}{8}$ inch for the difference between the nominal and actual thicknesses of the block.

Pile dry blocks near the corners, such as *A* and *B* in Figure 1, where they will be handy for use as the mortar is applied to the footing.

Have a trowel, a folding rule, and a mason's level ready for use.

Note the position of the first lead block, as shown in the illustration.

Mix the mortar and have it handy on a mortarboard. The board should be about 3 feet square. Trowel and turn the mortar (mix) several times as it is being used for block after block. This tends to temper the mortar and keep it plastic and workable.

Laying First Block of Lead (Corner *A*). Spread the bed-joint mortar as shown in the *Y* part of Figure 9. Use both hands to lift the first block

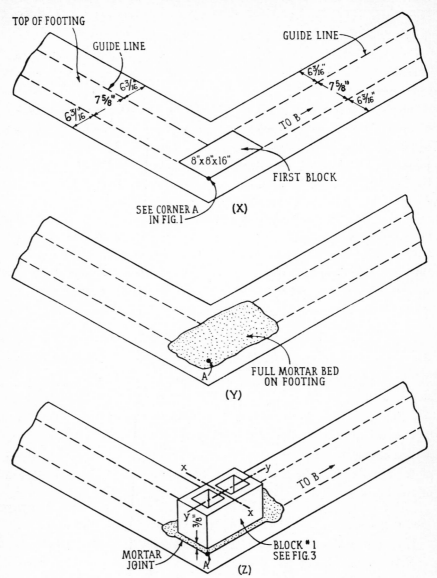

FIGURE 9. Laying first block in side *AB* shown in Figure 1.

up and then gently place it, as shown in the Z part of the illustration. Press down on the block and remove the excess mortar with the edge of the trowel. NOTE: Do not *scrape* the trowel along the block—instead, use the edge of the trowel to *cut* the mortar, as with a knife. Measure the joint thickness. Press the block down until the thickness is ⅜ inch. It is permissible to use the handle or blade of the trowel as a means of pounding blocks down into place. If the joint should become too thin, remove the block, apply more mortar, and try again.

After the joint thickness is correct, place the level as shown by the *x-x* and *y-y* dashed lines. If necessary, tap the top of the block with the handle of the trowel until the block is perfectly level in both directions. Then measure to make sure that the end and long side of the block are exactly 6¾₁₆ inches from the edge of the footing.

If the level and position of the block do not check properly, do not hesitate to remove the block and try again. This is the *practice* inexperienced masons need. Getting that practice may take some time and effort, but it is not a sin to keep trying until the block is perfectly laid.

Trim up the mortar joint under the block. Make it look neat and attractive. Then, once the block is all set, take care not to disturb it.

Laying Second Block of Lead. Spread the bed mortar for the second block of the lead as shown in the X part of Figure 10. Butter one end of the second block, as shown in Figure 7. Lift the block with both hands and gently place it in position, as shown in Figure 11. Very gently place the block against the first block and press it down into the mortar joint until it appears as shown in the Y part of Figure 10. Remove excess mortar, and check the thickness of the horizontal and vertical joints. Check the level as indicated at *x-x* and *y-y*. Check the position of the block to be sure it is exactly 6¾₁₆ inches from the edge of the footing.

Look at the mortar joints thus far made. If they are not full of mortar, use the point of the trowel to press more mortar into them. The joint at *t* can be filled, too. Be sure that the joint mortar is flush with the faces of the block.

Laying Third Block of Lead. Spread the mortar as shown in the Y part of Figure 10. Lay the block as explained for the second block. Check

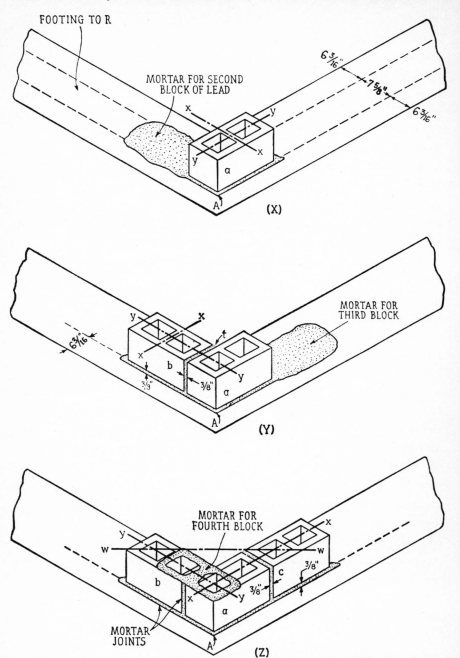

FIGURE 10. Laying second and third blocks in side *AB* in Figure 1.

levels by placing the level in the positions of dashed lines *x-x*, *y-y*, and *w-w*, as shown in the *Z* part of the illustration. Do not slight the level checking. It is the most important part of the work.

Laying Fourth Block of Lead. Spread the mortar for the fourth block as shown at *Z* in the illustration and as shown in Figure 8. Place the block as shown in the *X* part of Figure 12. Use the level in the positions of dashed lines *x-x*, *y-y*, *u-u*, and *z-z*. The exterior faces of this block must be in vertical alignment with blocks *a* and *b* under it.

FIGURE 11. How to set block into position. (Courtesy of Portland Cement Association.)

Finished Lead. See the *Y* part of Figure 12. Blocks *e* through *i* are laid in the same manner. The lead for the first three courses, at corner *A*, will then be complete. Use the level in the positions of dashed lines *y-y*, *w-w*, *t-t*, *u-u*, and *v-v*. Every such level must be *perfect*. Otherwise, tear down the lead and try again. If all horizontal joints were made ⅜ inch thick, the dimensions should be 24 inches, because each block is 7⅝ inches high and there are three of the ⅜-inch joints.

Lead for Corner B in Figure 1. The lead for the corner at *B* is laid up following exactly the same procedure as explained previously. Figure 13 shows the leads at corners *A* and *B*.

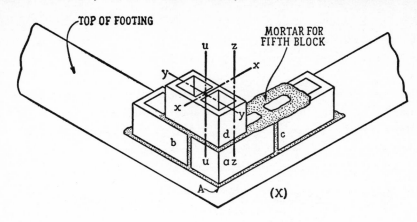

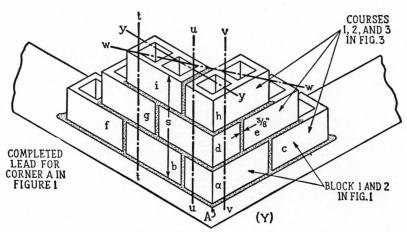

FIGURE 12. Lead for corner *A* in Figure 1.

Laying Block Between Leads. Figure 13 shows the footing for side *AB* of Figure 1. Both leads for the *A* and *B* corners are complete.

Note the nails at the *A* and *B* corners. Between these nails, a stout string or cord should be tightly stretched. It should lie exactly along the top edge of the block in the first course of the leads.

The mortar for the first block, between leads in the lower course, is placed as previously described. The end of the block should be buttered as shown in Figure 7 and the block laid as explained above. The cord is

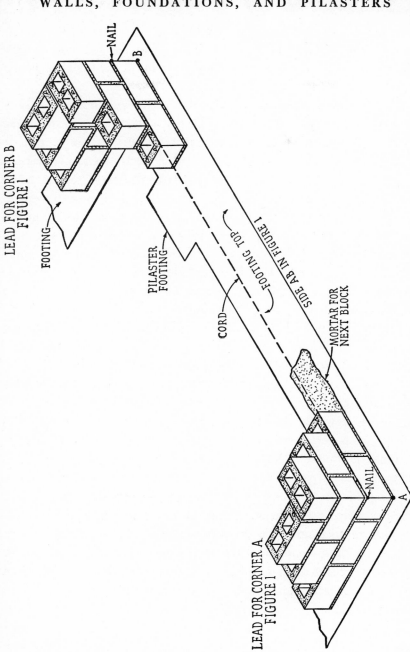

FIGURE 13. Leads for wall *AB* in Figure 1.

used as a guide, and the top outer edge of the block should be in line with it, as shown, where the cord touches the lower-course block in the two leads. This procedure will automatically control the thickness of the bed joint.

Concrete blocks are not always *exactly* 15⅝ or 7⅝ inches long. For this reason, it is necessary to place dry block (without mortar) along the footing between leads to determine if the vertical joints can be exactly ⅜ inch or if they will have to be increased or decreased to compensate for variations in block lengths. Whatever variation exists can be taken care of by slightly increasing or decreasing the usual ⅜-inch vertical joints. For example, if the blocks are somewhat short, the seven joints required between leads may take up 1 inch or more. In such a case, each of the joint thicknesses can be increased a trifle so that the inch is equally distributed among them. This is one of the procedures which masons have to work out according to the requirements of the blocks they are using. However, once any block inaccuracies are noted, they can be compensated for without much trouble. It is just a case of remembering to do it.

Other first-course block are laid between leads in the same manner.

For the pilaster, an additional block is laid to make the required 16-inch-square area.

Figure 14 shows the first course, including the pilaster, complete between the leads for corners *A* and *B*. At this stage, the cord must be moved up one course, as shown.

Figure 15 shows part of the second course, including the pilaster, between the leads for corners *A* and *B*. The three courses between the leads can be completed using the same general procedures already explained. Care must be taken to check the vertical alignment of each block laid. Place the level in the position of the dashed line *x-x* in Figure 15 as each block is laid. This is the only method of keeping the interior faces of all blocks in a wall or foundation and the whole wall or foundation in the same vertical plane.

Leads at Corners C, B, etc. Once the side *AB* in Figure 1 has been laid up three courses high, leads at *C, D,* etc., should be laid up and

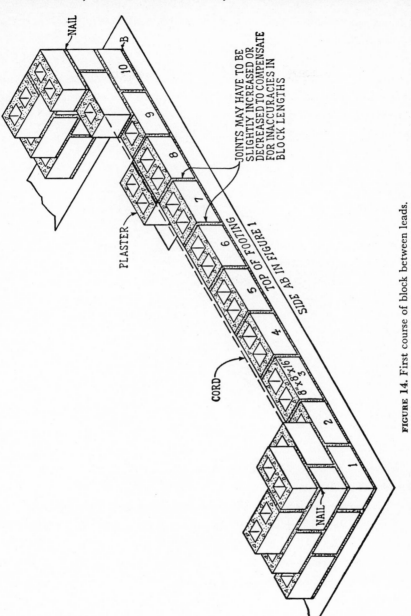

FIGURE 14. First course of block between leads.

then the intervening block, until the whole wall or foundation is three courses high.

Courses 4, 5, and 6 (Side *AB*). For the next three courses, three-course leads are again laid up at corners *A* and *B* and the previously explained procedures repeated all the way around the foundation. In the pilaster, the blocks are alternated as shown in Figures 14 and 15.

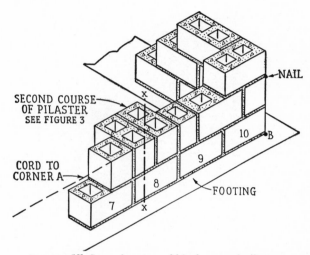

FIGURE 15. Second course of block around pilaster.

Courses 7 through 10 (Side *AB*). As shown in Figure 16, these courses include the window opening. Offset jamb blocks, as shown in Figure 18 of Chapter 9, Typical Concrete-block Details, are required. This opening can be laid up without having the required metal window frame set up.

Before the jamb blocks are laid up, the actual measurements of the steel window frame should be known so that the opening can be made to fit it exactly. Some variation in block lengths and heights may have to be compensated for, in both horizontal and vertical joints, all around the window opening. Such variations are not generally sizable, and joint thicknesses can easily be controlled to compensate for them. This is another of the variables which masons must keep in mind.

If possible, the sill should be precast in order to be sure that its slope will exactly fit the sill portion of the steel window frame. Note that sills should be laid on a full mortar bed, as shown at Y, and that the joint should be the same thickness as the horizontal joints between the blocks at either end of it.

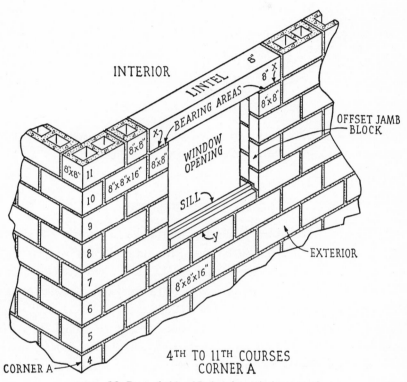

FIGURE 16. Part of side AB showing window opening.

Course 11 (Side AB). This course, as shown in Figure 16, includes the concrete lintel above the window opening. Note that the bearing areas must be at least 8 inches long, as shown at x, and that a full mortar joint is necessary at each such area.

The lintels can be precast, cast on the job, or cast in place.

Top Course (Side AB). Figure 17 shows how the top course, which consists of $3\frac{5}{8}$-inch solid concrete blocks, is laid up. Such blocks are laid

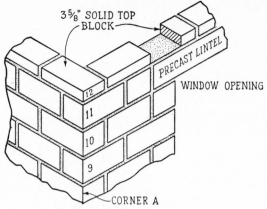

FIGURE 17. Top course for foundation.

exactly as regular blocks and with the same joint thicknesses. They should conform to the regular block pattern of the foundation.

JOINT TOOLING

As explained in Chapter 3, Mortar, mortar has a tendency to shrink a little as it hardens. When this happens, the mortar pulls away from the edges of the block and creates a thin crack. Such a crack could account for the lack of watertightness in a wall or foundation. In order to avoid such a possibility, the joints should be tooled.

There is still another reason why joints should be tooled. When the excess mortar is scraped off joints at the time the blocks are laid, some of the mortar may be pulled away from the edges of the blocks, causing a crack. The tooling avoids that kind of crack, too.

When joint mortar has stiffened somewhat, it can be firmly packed by the use of a jointing tool, such as shown in Figure 18. This tool compacts the mortar in both horizontal and vertical joints. Pressing this tool against the mortar in joints and then sliding the tool along restores contact between the mortar and the block and helps materially to make weathertight joints. It may be necessary to add a little mortar, particularly to vertical joints, to ensure that they are well filled. After the joints have been tooled, any excess mortar on either side of the joints can be removed by the edge of the trowel.

FIGURE 18. Tooling mortar joint. (Courtesy of Portland Cement Association.)

EXAMPLE 2   (Reinforced)

For this example, Figures 1 through 5 can be assumed as posing the general problem, except that steel reinforcement is to be added.

As previously set forth, various building codes differ in the amount and placing of steel required. In order to include possibly the greatest requirement for reinforcement, this example makes liberal use of steel. Readers may satisfy the requirements of their own localities by using less steel wherever allowable. However, the amount of steel used in this example is considered the best practice.

Planning Reinforcement. As explained in Chapter 9, Typical Concrete-block Details, dowels are generally spaced about 24 inches apart and must extend 24 inches above the surface of a footing. However, because of pilasters and actual dimensions of any given project, exact spacing of 24 inches is not always possible. Therefore, careful planning must be done prior to the time the concrete for the footings is placed. This is another case where scaled drawings serve an excellent purpose.

The upper part of Figure 19 shows a top view of the ten blocks required for the first course of the section or side of the foundation (it could just as well be a wall entirely above grade) shown between A and B in Figure 1. Including mortar joints, each block is 16 inches long. The problem is to plan the dowel spacing so that they will be exactly in the cores of the block.

Two dowels are generally required at each corner, such as A and B, and on both sides of window and door openings. Thus, blocks 1, 2, 5, and 10 will each have two dowels, as indicated by the dots, in their cores. This can be visualized to better advantage by studying Figure 3 in which

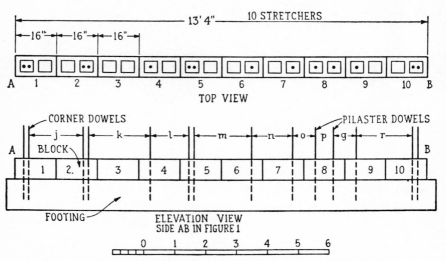

FIGURE 19. Dowel-location planning for side AB in Figure 1.

the circled numerals indicate the position and number of dowels in blocks 1, 2, 5, and 10.

Four dowels, one for each core, are generally required for pilasters, as shown in Figures 14 and 15. Again referring to Figure 3, it can be seen that the pilaster dowels will appear in block 8. The dots indicate the dowels in the top view of Figure 19.

In order to make the dowel spacing somewhere near the desired 24 inches, they are shown in blocks 4, 6, 7, and 9, as indicated in the top view of Figure 19.

The elevation view of Figure 19 shows the blocks, as seen from one side of the foundation, and the required dowels. Inexperienced masons should make such a drawing and then assign actual dimensions to the spacings marked *j* through *r*. Then when the footings are being placed, the dowels can be spaced accurately, and they will be in the proper cores of the blocks. This is an important aspect of layout and should be carried out for all sides of the walls or foundations. (See Chapter 14, Illustrative Example, for a method of determining dowel-spacing dimensions.)

Studs (reinforcement above the footings to the full heights of walls and foundations) should be in the same locations as the dowels and lap them at least 24 inches. Thus, the dots in the upper part of Figure 19 also indicate the required studs.

The *X, Y,* and *Z* parts of Figure 20 indicate the dowel positions in, and on both sides of, the first block of the lead at corner *A,* shown in Figure 1. All the preparation and laying procedures are the same as explained for Figure 9 in Example 1.

The *X, Y,* and *Z* parts of Figure 21 show how the dowels should be located in the cores of the first, second, and third blocks of the lead at corner *A.* The block-laying procedures are exactly the same as for Figure 10 in Example 1.

The *X* part of Figure 22 shows the dowels for the fourth block of the lead for corner *A.* The *Y* part of the illustration shows the complete lead. Note that in blocks *g* and *h* the dowel tops are at the same level as the tops of the blocks. The block-laying procedures are exactly the same as for Figure 12 in Example 1.

Figure 23 shows the dowels (including the four necessary for the pilaster) required in side *AB.* They check with the dowel indications shown in Figure 19. The block-laying procedures are exactly the same as for Figure 13 in Example 1.

Figure 24 shows the entire first course of block between corners *A* and *B* and the dowels in the proper cores.

Figure 25 shows a portion of the second course of blocks between corners *A* and *B* and the pilaster dowels in their proper cores.

Bond Beams. Figure 26 shows the use of a reinforced 3⅝-inch bond beam at the top of the foundation (or wall). A U-shaped block, with open ends down, can be used. Note that the bond block is mortared to other blocks by the usual mortar joints.

The steel rods in the bond beam are first placed and the beam blocks

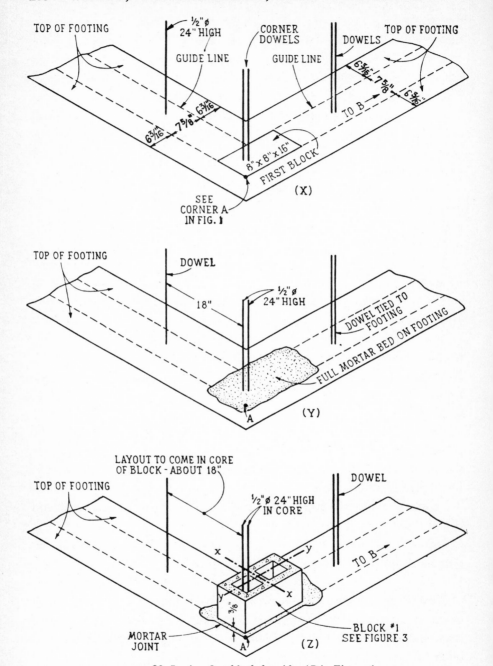

FIGURE 20. Laying first block for side *AB* in Figure 1.

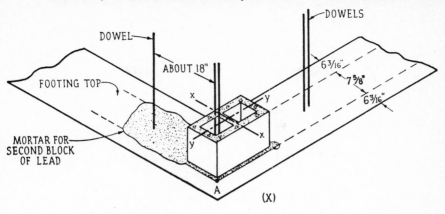

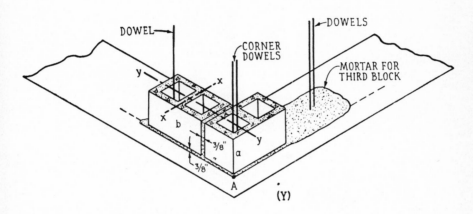

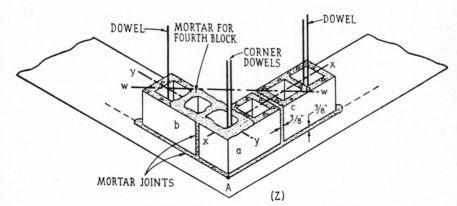

FIGURE 21. Laying second and third blocks for side *AB* in Figure 1.

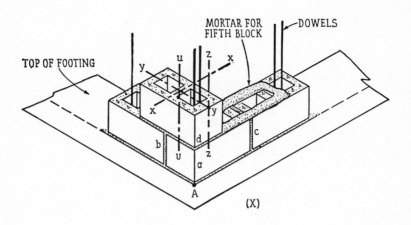

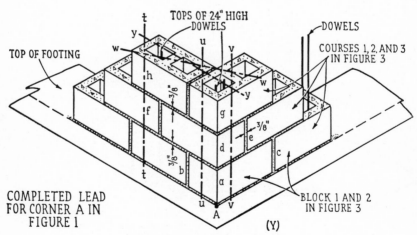

FIGURE 22. Lead for corner *A* in Figure 1.

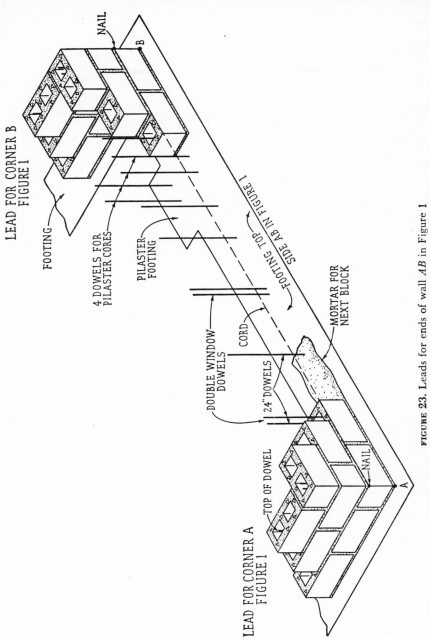

LEAD FOR CORNER B
FIGURE 1

NAIL

B

FOOTING

4 DOWELS FOR
PILASTER CORES

PILASTER
FOOTING

FOOTING TOP IN FIGURE 1

SIDE AB IN FIGURE 1

DOUBLE WINDOW
DOWELS

CORD

24" DOWELS

MORTAR FOR
NEXT BLOCK

TOP OF DOWEL

NAIL

A

LEAD FOR CORNER A
FIGURE 1

FIGURE 23. Leads for ends of wall *AB* in Figure 1

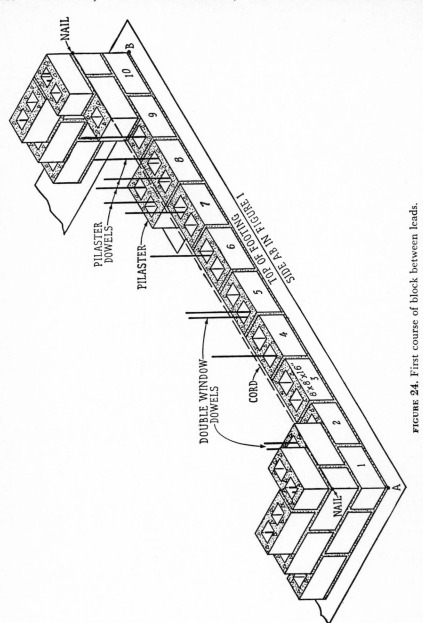

FIGURE 24. First course of block between leads.

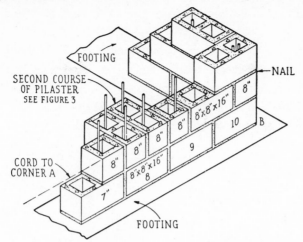

FIGURE 25. Second course of block around pilaster.

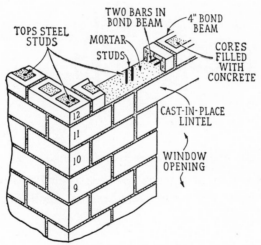

FIGURE 26. Four-inch bond beam.

laid on top of them. Finally, pebble (or pea) concrete is used to fill all stud cores, down to the surface of the footing and flush with the top of the bond beam. The foundations (or walls) are thus tied together by several reinforced-concrete columns and the reinforced bond beam. As the core concrete is placed, it should be carefully spaded, using a long piece of reinforcing rod, so that the concrete completely fills all cores. The placing of core and bond-beam concrete should be done all in one operation so that the bond beam and all vertical columns (studs) are integral.

The lintel over the window opening must be cast in place so that its reinforcement and the stud reinforcement, on both sides of the window opening, will be in integral concrete.

Figure 27 shows a 16-inch bond beam. The general construction procedures are the same, except that, when a window opening is directly

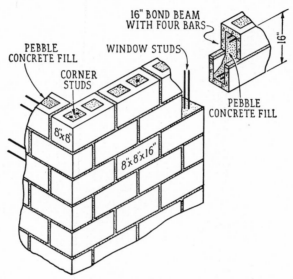

FIGURE 27. Sixteen-inch bond beam over window opening and at top of wall.

under such a beam, the beam also serves as a lintel and temporary supports are necessary across the window span until the concrete has hardened.

TUCK POINTING

Sometimes the mortar joints of existing walls and foundations, because of original careless workmanship or natural weathering, crumble and fall out to the extent that the joints are no longer water or weatherproof. Such a condition can be repaired by means of *tuck pointing*.

All old mortar should be scraped out of the joints to a depth of at least 1 inch. An old screwdriver or a cold chisel and a hammer can be used for the purpose. All loose mortar, at whatever depth in joints, should be removed. All traces of dirt and loose mortar particles should be brushed out of the joints.

Ordinary new mortar can be used for the repair work. It should be somewhat stiff so that it can be pushed into the joints using a tuck-pointing trowel. Moisten the interiors of the joints and apply the new mortar in three layers. Roughen the surface of the first two layers so that succeeding layers will bond or stick tightly. The new mortar should be pressed firmly into the joints as a means of making sure that they are completely filled. The final layer should be flush with the exterior faces of the block.

After the mortar of the final layer has stiffened somewhat, the joints should be tooled as previously explained in this chapter.

CHAPTER ELEVEN

Concrete-block Chimneys

Both solid (*without* cores) and ordinary concrete blocks constitute excellent, economical, and generally approved materials for the construction of common chimneys in houses and other small buildings. When planning is well done and the workmanship good, chimneys built of such materials can be depended upon so far as fire safety, durability, appearance, and service are concerned.

The purpose of this chapter is to explain the important aspects of chimney planning and to suggest construction procedures.

PRODUCTS OF COMBUSTION

When various fuels, such as coal, gas, and oil, are used to create heat, the combustion (burning) processes give off what are known as *products of combustion*. Each fuel gives off such products in somewhat different form.

Smoke. When coal and oil burn, the combustion processes generally give off what is commonly known as *smoke*. The smoke is apt to contain unconsumed fuel, such as soot, and one or more gases which, besides being dirty and having an objectionable odor, may be dangerous to life and health. Some of the gases are odorless and invisible, but dangerous just the same.

244

Water Vapor. When fuel gas burns, the combustion process may also give off products which are dangerous to life and health. In addition, a considerable amount of water, in the form of an invisible vapor, is created.

PURPOSE OF CHIMNEYS

No matter what type of heating equipment is employed and no matter what kind of fuel is burned, chimneys should be planned to serve two major purposes.

First Purpose. As previously explained, smoke and various associated gases are objectionable and dangerous. Thus, the first purpose of chimneys is the proper disposal of such products. It was also pointed out that the burning of fuel gas produces large quantities of water vapor. Chimneys must dispose of such vapor in order to avoid excessive humidity and the resultant possibility of damage to structural details and decorations.

Second Purpose. In order for combustion processes to take place, with any kind of fuel, ample amounts of oxygen must be present. Since natural oxygen is present in the air, it can be readily understood that fuel beds and burners must be supplied with a constant supply of air.

As the burning process of any type of fuel takes place, the products of combustion are warm and rise because of the decrease in density owing to heat expansion. As they rise through the interior of the chimney, a *draft* is created because new air is drawn in through the firebox or burning compartment. Thus, the continued draft keeps an ample supply of air (oxygen) where it is needed.

The second purpose of chimneys is, therefore, to produce drafts which, in turn, cause the required amount of air to circulate through the areas where combustion takes place.

KINDS OF CHIMNEYS

Generally speaking, chimneys are classified, first, according to the materials used in their construction and, second, according to the type of heating equipment they serve. In this book, only concrete-block chimneys are considered.

Coal- and Oil-fired Equipment. For coal- and oil-fired heating equipment, what are known as *universal* chimneys may be used. The flues in such chimneys are large in order to dispose of the products of combustion properly and to create ample draft. If such chimneys are properly planned and constructed, they can serve the needs of *any* equipment. In fact, universal chimneys often are a good choice because they are ready to serve any possible change in equipment without having to be remodeled or rebuilt.

Gas-fired Equipment. When chimneys are planned and built especially to serve the needs of gas-fired heating equipment, their flues may be much smaller than are required in universal chimneys. There is appreciable economy, so far as labor and materials are concerned, and such chimneys do not require much space.

CHIMNEY LOCATIONS

It was previously pointed out that *draft* is important for the burning of fuels. The best draft possibilities exist when the chimneys are located *within* buildings *away from* exterior walls.

As the products of combustion leave their source, they are warm and tend to rise. They continue to rise, up and through chimneys, so long as they stay warm. However, if they become chilled, their tendency to rise stops, and as a result draft is seriously affected.

When chimneys are located within buildings, their bulks, or masonry masses, become warm. Then, the products of combustion rise through them without being chilled. On the other hand, if chimneys are located in cold exterior walls, their masonry masses become cold, and the products of combustion are chilled.

Thus, wherever architectural layouts and structural details permit, chimneys should be so planned that they can be located *within* buildings. In cases where interior locations are not possible, the sides of chimneys exposed to low temperatures should be at least 8 inches thick. The added thickness will help to keep the flue areas warmer and make better draft possible.

PURPOSE OF FLUE LINING

Flues are the interior areas or parts of chimneys through which the products of combustion pass. Typical examples are shown in Figures 3 and 5.

The flues in all chimneys, regardless of the type of heating equipment served, should be lined with materials especially made for that purpose. There are several good reasons why linings are necessary.

Fire Safety. In spite of frequent cleaning, the flues in chimneys which serve coal- and oil-burning heating equipment become coated with deposits of soot. This is unavoidable. In many cases, the soot deposits become thick.

Soot will burn! It often does so and at unpredictable times. When it burns, it creates exceptionally high temperatures and pressure in the flues. Such pressure is not readily relieved through the opening at the top of the chimney and tends to force fire through any cracks.

Even though the mortar joints in concrete-block or any other kind of masonry chimney are full and carefully tooled, there is always the chance that the pressure built up in the flues will force fire through them and endanger surrounding wood or other combustible materials. Many disastrous fires have been caused in this manner. Good flue linings absolutely prevent such danger.

Cleaning. All chimneys will function to better advantage when their flues are clean and smooth. A clean flue offers less resistance to the passage of the products of combustion. Where flue linings are used, the cleaning can be done with much less effort.

Fireplaces. When wood is burned in fireplaces, certain volatiles, such as sap, resin, and other gums, are vaporized and become part of the products of combustion. Within the flues of chimneys serving fireplaces, the vapor may be sufficiently cooled to condense. When this happens, a dark creosotelike liquid is deposited on the sides of the flues. In time, the liquid may soak through mortar joints and ruin the appearance of plaster and other adjacent structural details. Where flue linings are used, such a possibility is impossible.

Gas Furnaces. As previously explained, water vapor is given off when fuel gas burns. Unless flue linings are used, that water will soak through mortar joints and run down the sides of chimneys, where it can cause a great deal of damage to other structural details.

KINDS OF FLUE LINING

Most flue linings are made of either terra cotta or glazed clay at least ⅝-inch thick. Figure 1 shows the shapes most commonly employed.

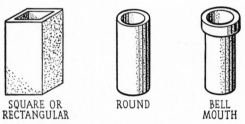

SQUARE OR   ROUND   BELL
RECTANGULAR   MOUTH

FIGURE 1. Typical kinds of flue lining.

Square, rectangular, or round flue linings may be used in universal chimneys. Some schools of thought hold that the round linings cause less resistance to the passage of the products of combustion.

Bellmouth linings are almost invariably used when chimneys are planned and built for use with gas-fired heating equipment. The glazed surfaces and bell connections resist the escape of water vapor.

NUMBER OF FLUES PER CHIMNEY

Each source of combustion should have a separate flue. In other words, if a house has a furnace and a fireplace, *two* flues are necessary. This is an important rule because when two sources of combustion are attached to just one flue, neither source can function properly. When gas hot-water heaters are employed, each should have a separate flue.

HEIGHTS OF CHIMNEYS

The height of a chimney is generally controlled by the type of building. Whether in one- or two-story houses or other buildings, the top of the chimney should be far enough above the roof so that air currents will not be deflected down into the flues. If this happened, the draft would be stopped. Figure 2 shows the approved recommendations.

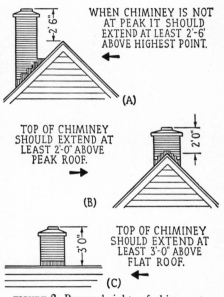

WHEN CHIMINEY IS NOT AT PEAK IT SHOULD EXTEND AT LEAST 2'-6' ABOVE HIGHEST POINT.

(A)

TOP OF CHIMINEY SHOULD EXTEND AT LEAST 2'-0" ABOVE PEAK ROOF.

(B)

TOP OF CHIMINEY SHOULD EXTEND AT LEAST 3'-0" ABOVE FLAT ROOF.

(C)

FIGURE 2. Proper heights of chimney tops above roofs.

FIGURE 2 (A). When the chimney is not at the peak of the roof, its top should extend at least 2' 6" above the peak or highest point of the roof.

FIGURE 2 (B). When the chimney is at the peak of the roof, its top should be at least 2' 0" above the peak.

FIGURE 2 (C). In regard to flat roofs, the top of the chimney should be at least 3' 0" above the deck or roof.

CHIMNEY BLOCK

Figure 3 shows a group of typical chimney blocks of the kind often used for ordinary one-, two-, and three-flue chimneys.

FIGURE 3 (U). The blocks in details A through I are solid (without cores). There are three basic sizes as shown in the Y part of the illustration. Both modular and nonmodular (standard) sizes are shown at Y.

FIGURE 3 (V). This type of block, also in modular and nonmodular

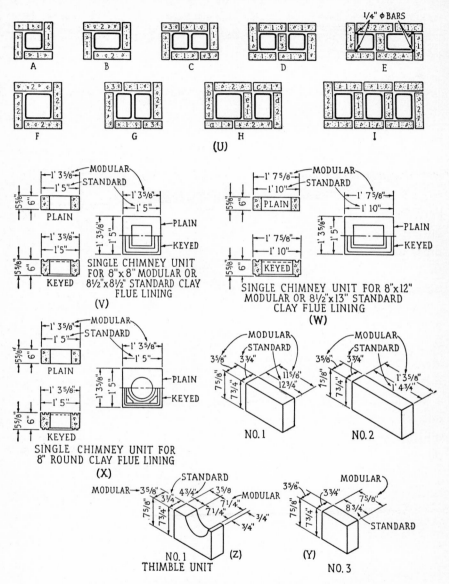

NOTE:- THE CONSTRUCTION DIAGRAMS SHOWN CONFORM
TO THE BASIC REQUIREMENTS OF THE CHIMNEY
CODE OF THE NATIONAL BOARD OF FIRE
UNDERWRITERS

FIGURE 3. Modular- and nonmodular- (standard) size chimney block for use with flue linings.

sizes, is laid one above the other, with mortar joints between them, to form a flue of 8- by 8-inch (modular) and 8½- by 8½-inch (nonmodular) dimensions. Note that plain or keyed blocks are available.

FIGURE 3 (W). This type, similar to the type shown at V, is for a somewhat larger flue lining.

FIGURE 3 (X). This type, also modular and nonmodular in size, allows for an 8-inch round flue lining.

FIGURE 3 (Z). Two of these blocks can be substituted for two of the No. 2 block, shown at U and Y, to form a thimble opening for a smoke pipe.

If modular block of the ordinary kind, such as shown in Figure 1 of Chapter 8, Concrete Block, and general modular dimensions are used for a house or other building, then modular chimney block should be used. Otherwise, the nonmodular chimney block should be used.

TABLE 1. Concrete chimney units required for different sizes of flue linings shown in Figure 3

Detail	Number of flues	Number of linings	Nominal dimensions of clay flue linings, inches		Solid concrete-block units required per course		
			Modular	Nonmodular	No. 1	No. 2	No. 3
A	1	1	8 × 8	8½ × 8½	4		
B	1	1	8 × 12	8½ × 13	2	2	
C	2	2	8 × 8	8½ × 8½	4	..	2
D	2	2	8 × 8	8½ × 8½	6		
E	2	1	8 × 8	8½ × 8½	4	2	1
		1	8 × 12	8½ × 13			
F	1	1	12 × 12	13 × 13	..	4	
G	2	2	8 × 12	8½ × 13	2	2	2
H	2	1	8 × 12	8½ × 13	3	4	
		1	12 × 12	13 × 13			
I	3	3	8 × 12	8½ × 13	8	2	

Table 1 shows the number of chimney units required per course for the different sizes of flue linings shown in the U part of Figure 3. For example, the detail shown at B has one flue lining. If the flue is to be 8

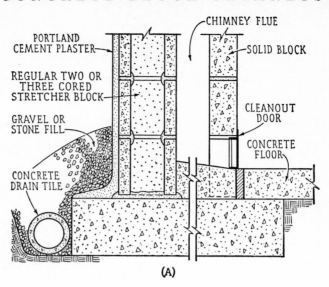

(A)

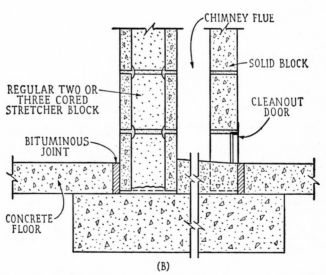

(B)

FIGURE 4. Use of regular concrete block in chimney construction.

by 12 inches in size, two No. 1 and two No. 2 block are necessary for each course.

There are many other styles and shapes of chimney block available from various manufacturers, any of which can be used as well as the typical block shown in Figure 3.

Regular two- and three-cored stretcher blocks can also be used in chimney construction. The *A* and *B* parts of Figure 4 show regular stretchers used for a chimney in an exterior wall and for a chimney which is part of an interior wall. The solid block indicated is of the kind shown at *L* in Figure 1 of Chapter 8, Concrete Block.

When one-flue chimneys are to be built, using block such as shown at *V* and *W* in Figure 3, it is best to use the keyed block when the chimneys will be exposed to the weather.

GENERAL PLANNING ASPECTS

Before starting to plan any concrete-block chimney, several general planning aspects should be understood and then kept in mind throughout the planning and construction.

Reinforcement. Concrete-block chimneys, such as shown at *E* in the *U* part of Figure 3, can be reinforced by placing ¼-inch or larger steel rods as indicated. Such rods are embedded in the mortar between the units and the flue lining and give the chimneys a great deal of additional strength with which to resist wind pressure and earthquake shocks. The rods must each be in one continuous piece for best results. Readers should check their local building codes before planning reinforced chimneys.

Fireproofing. All wooden floors, walls, roofs, etc., must be framed around chimneys in such a manner as to create a 2-inch space between them and the chimneys. Such spaces should be filled with an incombustible insulating material. Figures 5 and 6 show such details.

Flashing. Typical flashing is indicated in Figure 6. Note that reinforcement is necessary for stucco over the flashing.

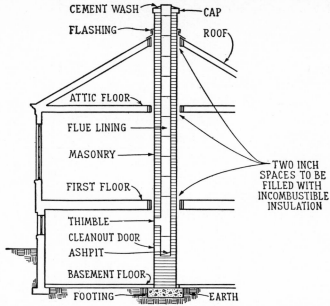

CEMENT WASH — — CAP
FLASHING — — ROOF
ATTIC FLOOR
FLUE LINING
MASONRY — — TWO INCH SPACES TO BE FILLED WITH INCOMBUSTIBLE INSULATION
FIRST FLOOR
THIMBLE
CLEANOUT DOOR
ASHPIT
BASEMENT FLOOR
FOOTING — — EARTH

FIGURE 5. Details of necessary fire-safety spaces between chimney and other structural details.

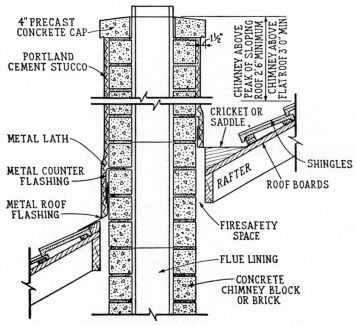

4" PRECAST CONCRETE CAP
PORTLAND CEMENT STUCCO
METAL LATH
METAL COUNTER FLASHING
METAL ROOF FLASHING

CHIMNEY ABOVE PEAK OF SLOPING ROOF 2'6" MINIMUM
CHIMNEY ABOVE FLAT ROOF 3'0" MIN.
1½"
CRICKET OR SADDLE
SHINGLES
RAFTER
ROOF BOARDS
FIRESAFETY SPACE
FLUE LINING
CONCRETE CHIMNEY BLOCK OR BRICK

FIGURE 6. Typical details of block chimney at roof line.

Cricket or Saddle. Figure 6 illustrates how a cricket or saddle can be used on the high sloping side of a roof to shed water around chimneys.

Footing. Chimneys should be supported by placed concrete footings which are located below frost line. Poor footings may result in tilting, uneven settlement, and dangerous cracks in chimneys.

Free-standing. Concrete-block chimneys should be free-standing. In other words, they should not require lateral support from other structural details.

Caps. Caps, such as shown in Figure 6, will shed water and preserve masonry below them. Flue linings should extend through caps and project a little above them.

Offsets. Chimneys should be built as nearly vertical as possible. However, offsets of not more than a few inches are allowable if the full areas of flues can be maintained at the offsets.

PLANNING CHIMNEY FOOTINGS

Chimneys, like all other structural details, require firm support. In fact, because they are relatively high details of small cross-sectional area, their supports (footings) must be especially firm in order to prevent tilting or leaning, settlement, or cracking.

Chimneys in Exterior Walls. When chimneys must be built as parts of exterior walls, their footings, as shown by Figure 6 of Chapter 5, Concrete Footings, must be *integral* sections of the footings for the walls or foundations.

Interior Chimneys. When chimneys can be built within a house or other building, they must have independent footings somewhat similar to those required for columns.

ONE-STORY BUILDINGS. Chimneys of one-story height should have placed-concrete footings which are at least 8 inches thick and projections of at least 4 inches on each side of the chimney.

TWO-STORY BUILDINGS. Chimneys of two-story height should have placed concrete footings which are at least 12 inches thick and projections of at least 6 inches on each side of the chimney. The use of steel reinforcing rods is recommended.

Soil Condition. The foregoing footing dimensions are for average two-flue chimneys and for soils having normal strength in terms of load-carrying capacity. When weak soils, which cannot safely support more than 2,000 pounds per square inch, must be considered, some rough calculations should be made in order to be sure that footing areas are ample.

Example. An ordinary two-flue fireplace chimney, of average two-story height, can be assumed to weigh about 40,000 pounds. This weight is probably more than that of the average chimney but constitutes a factor of safety in the planning of footings. Dividing 40,000 by 2,000 gives 20, which means that the footing, when placed on soft clay, must have an area of 20 square feet. In other words, the footing must be approximately 4' 4" square.

The chimney footing indicated in Plate II has dimensions of 3' 8" by 5' 8" and is for a one-story house. The chimney weight is much less than 40,000 pounds, and there may be some question as to why such a large footing is specified.

The fireplace, plus its surrounding masonry work, has a large horizontal area. That whole area must be supported. Thus, the chimney structure under it must have the same over-all dimensions. The footing must have enough area to include the chimney structural area plus at least 6-inch projections. Thus, the footing has a much larger area (20.71 square inches) than actually required for a one-story house if only the chimney weight were considered. It is generally good policy to plan fireplace chimney footings of considerably greater area than the chimney weights actually require. This policy, while requiring several more cubic feet of concrete, makes sure that leaning or tilting, settlement, and cracking cannot possibly occur.

PLANNING FLUES FOR COAL AND OIL FUELS

For ordinary one- and two-story, five- and six-room houses which have central heating systems (hot-air furnaces and steam or hot-water boilers) 8- by 12-inch (modular) or 8½- by 13-inch (nonmodular) rectangular

TABLE 2. Sizes and application of flue lining

Usual Application	Outside diameter, inches		Inside diameter, inches	
	Rectangular, modular	Rectangular, nonmodular	Circular	Bellmouth preferred for gas fuel
Gas heating, equipment	4 × 8	4½ × 8½	6	6
	8 × 8	8½ × 8½	6	6
	4 × 12	4½ × 13	6	6 or 7
Stoves, ranges, and room	8 × 8	8½ × 8½	8	8
heaters	8 × 12	8½ × 13	8	8
Coal- and oil-fired	8 × 12	8½ × 13	10	
central heating	12 × 12	13 × 13	12	
systems				

flues, as shown Table 2, are of sufficient size. Ten-inch circular flues are also of sufficient size.

PLANNING FLUES IN FIREPLACES

Table 3 shows recommended rectangular flue sizes for small-, medium-, and large-sized fireplaces. Ordinary fireplaces, such as those planned for most houses, can be classed as medium-sized. Therefore, 12- by 12-inch (modular) or 13- by 13-inch (nonmodular) are of sufficient size.

When ordinary houses are planned to have both a furnace and a fireplace, the details shown at H in the U part of Figure 3 serve as a guide for chimney planning.

TABLE 3. Fireplace-opening size and flue size

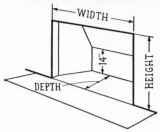

Size	Width, inches	Height, inches	Depth, inches	Flue lining, nonmodular	Flue lining, modular	Round, inches
Small	28	28	16	8½ × 13	12 × 12	10
	30	30	16	8½ × 13	12 × 12	10
Medium	34	30	18	13 × 13	12 × 12	12
	36	30	18	13 × 13	12 × 12	12
Large	40	30	18	13 × 13	12 × 12	12

PLANNING FLUES FOR GAS FUEL

For ordinary one- and two-story, five- and six-room houses which have central heating systems (hot-air furnaces and steam or hot-water boilers) rectangular, circular, or bellmouth flues may be used. In most cases, 8- by 8-inch (modular) or 8½- by 8½-inch (nonmodular) rectangular flues are of sufficient size. Also 6-inch circular or 6-inch bellmouth flues are of sufficient size.

When gas furnaces of the wall or floor types are to be used, 6-inch circular or bellmouth flues are of sufficient size. In some regions, the use of sheet-metal flues is allowed.

PLANNING FLUES FOR COAL-BURNING STOVES OR RANGES

Most heating stoves and cooking ranges can be amply served using 8- by 8-inch (modular), 8½- by 8½-inch (nonmodular), or 8-inch circular and bellmouth flues.

GENERAL FLUE PLANNING

In general, where space and material costs are not of great importance, it is the best policy to make flues somewhat larger than the minimum recommendations. This policy pays dividends in terms of ample draft and more certain disposal of undesirable smoke and gasses.

Flue and chimney sizes should be planned *before* final floor and other architectural plans are completed because the spaces they require, in terms of horizontal projection, are often greater than is realized. Wise planners go so far as to plan houses around fireplaces and chimneys. Then no serious difficulties can possibly confuse other planning aspects.

Each flue in any chimney should be entirely separate from other flues. There should be no cross connections between them. This rule should be followed wherever possible. However, because of space limitations, it is sometimes necessary to place two flue linings in the same flue space. While this sort of planning is to be avoided, the flue linings can be so placed that the joints of adjoining linings are staggered at least 7 inches.

Solid concrete block, as shown at *r* in the *D* detail in the *U* part of Figure 3, should separate all flue linings.

Flue linings should be laid ahead of the general construction of chimneys and the concrete block laid around the linings.

When flues must contain offsets, the offsets should, if possible, be put in fireplace flues. In other words, the flues for heating plants should be straight and vertical.

GENERAL CHIMNEY-BUILDING SUGGESTIONS

Chimneys should be laid up true and plumb and, unless otherwise specified, in running bond so that the joints are staggered in a manner similar to the way concrete-block walls and foundations are laid.

Mortar used in laying up block chimneys should be the same quality as used for walls and foundations. Mortar joints should be ⅜ inch thick with full beds under all solid block. Vertical joints should be full of mortar and tight.

Mortar may be placed in all spaces between flue linings and block.

This helps make chimneys strong and fire safe. In some cases, however, the mortar fill could create an expansion problem, leading to cracking or displacement of the lining or enclosure. For chimneys more than one story high, therefore, use of a thick liner and omission of the mortar fill would be preferable.

Flue linings should have mortar joints between their ends as they are laid in all chimney structures.

BUILDING A CONCRETE-BLOCK CHIMNEY

In order to illustrate the procedures recommended for inexperienced masons, it can be assumed that a concrete-block chimney, such as in-

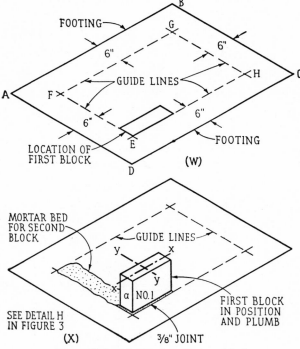

FIGURE 7. Starting to lay chimney block.

dicated at *H* in the *U* part of Figure 3, is to be constructed. The following suggestions also assume that the two flue linings start at the footing. This is somewhat apart from a typical case but will serve the purposes of this explanation.

It can be further assumed that the chimney footing is laid out and placed as explained in Chapter 5, Concrete Footings.

First Step. Assume the footing *ABCD,* as indicated in the *W* part of Figure 7. A 6-inch projection beyond the sides of the chimney is required. Therefore, the guide lines (which can be drawn using a pencil or chalk) are created as shown. Once drawn, the guide lines should be carefully checked to make sure they include the over-all dimensions of the chimney. Such dimensions can be visualized if dry block and flue linings are placed in the pattern shown by the *H* detail illustrated in the *U* part of Figure 3 and are then measured.

If the rough framing (joists) above the footing is in place, a plumb bob should be dropped from the corners of the opening in the framing (remembering the required 2-inch fireproofing) to make sure that the guide-line corners, shown at *EFGH* in the *W* part of Figure 7, are in exactly the correct positions. These corners will check out perfectly if the footing and framing were correctly located.

It is recommended that block *a* (see the *H* detail in the *U* part of Figure 3) be laid first. Its position is shown in the *W* part of Figure 7.

Second Step. Apply a full mortar bed, about 1 inch thick, and press the *a* block into position, as shown in the *X* part of Figure 7. Check the mortar joint to make sure it is exactly ⅜ inch thick. Plumb the block by placing a level in the position of dashed line *x-x* and *y-y*.

Apply mortar for block *b*. Also apply mortar to the end of block *b*, where it fits up against block *a*. Press the block gently into position, as shown in the *Y* part of Figure 8, and then plumb it and block *a* by placing a level in the position of the dashed lines *x-x, y-y,* and *z-z*.

Third Step. Place blocks *c* and *d*, as shown in the *Y* part of Figure 8, next, so that the first units of the flue lining can be set into position. Plumb block *c* and *d* as shown by the dashed lines *x-x, y-y,* and *z-z*. Then plumb all four blocks by placing a level at several positions, such as indicated by the dashed line *w-w*.

Fourth Step. Place units of flue lining and the block between them, as shown in the *Z* part of Figure 8. The flue-lining units and the block at *e* can be arranged so that mortar joints are all correct. Plumb the flue-lining units as shown by the dashed lines *x-x, y-y,* and *z-z*.

Fifth Step. Lay the remaining block to complete the first course and fill all spaces around the flue linings with mortar. Care must be taken so that the mortar completely fills all the spaces. It should be pressed down until no more can be added. Again check the flue linings to make sure that they are perfectly vertical.

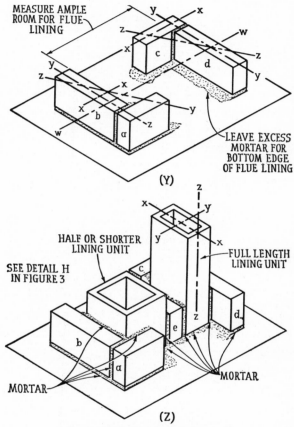

FIGURE 8. Placing first lining units.

Sixth Step. Place the second course of block and the next full unit of flue lining for the large flue. The courses should be staggered. Be sure that spaces around the flue-lining units are filled with mortar.

Seventh Step. Place the third course of block and be sure that all spaces are filled with mortar. Check the vertical alignment by placing the level as indicated by the dashed lines *x-x* and *y-y* in Figure 9. The level of each course should be checked as laid. This is extremely important because just a little inaccuracy can easily cause the whole chimney to be out of plumb when it gets to a height of several feet.

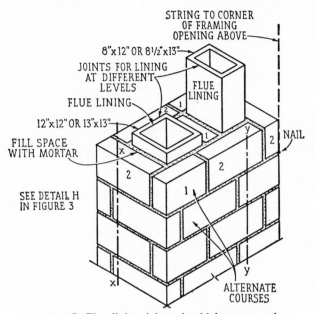

FIGURE 9. Flue-lining joints should be staggered.

Strings can be stretched from the corners of the framing openings, as indicated in Figure 9, and attached to the corners of the chimney. The strings will help to keep the corners in proper vertical alignment as the other courses are laid.

CHAPTER TWELVE

Stucco

In Southern, Southwestern, and, to some extent, Eastern regions of the country, stucco exterior finishes provide ample protection from the elements and beautiful architectural effects so far as houses and other small buildings are concerned. When properly applied, using the recommended materials, such finishes are durable and will not crack. Stucco is well suited to either frame or masonry construction, can be colored to suit any taste, and given any one of several textures.

Good materials and workmanship are essential in all types of construction. However, they are especially important in the application of stucco, used as it is in a relatively thin concrete shell on the exterior of buildings where it must withstand exposure. The purpose of this chapter is to illustrate and explain the more important factors which must be understood and applied in order to obtain successful stucco work.

AGGREGATE

Aggregate for stucco should be composed of natural sand which is clean and well graded, as explained in Chapter 4, Concrete. Such sand can be secured from dealers, or it can be cleaned and graded as also explained in Chapter 4. Dirty sand is especially harmful to stucco work and should be avoided at all costs.

The use of coarse, well-graded sand improves the workability of stucco, increases its durability, and lessens the possibility of crazing or severe

264

cracking. This is one of the recommendations concerning material to which inexperienced masons should pay particular attention. So-called *concrete* sand should be used for all stucco work. If such sand should lack sufficient fine material (which is necessary in making a mix more workable), the proper grading can be obtained by adding limited amounts of so-called *plastering* sand.

CURING

As with all masonry construction where mixed materials, such as mortar or concrete, are being applied, curing deserves special attention. It is essential that the stucco be kept damp until setting and hardening have taken place. Proper moist curing helps to develop maximum strength and density, reduces the possibility of troublesome shrinkage, and avoids crazing and unsightly cracks.

As each coat of stucco is applied, it should be kept moist over a period of 24 to 48 hours by means of a spray of water. The surface should not be flooded, as though by rain, but just wetted enough to maintain a moist condition. This is a tedious chore, but it pays great dividends in any stucco work. Its importance cannot be stressed too much. In fact, curing is every bit as important as materials and proper application procedures. Inexperienced masons will do well to keep curing in mind.

BOND

In order to be sure that stucco will properly adhere (bond), the surface to which it is applied must have a certain degree of suction. The suction draws the stucco mortar into the pores of the base or previously applied coat of stucco where it can get a good grip, so as to speak. Uniform suction is obtained by controlling the rate and amount of absorption. This is done by dampening (not soaking) the surface, such as the concrete block or the previous coat of stucco, to which the stucco is to be applied. This, too, is a tedious chore, but it pays dividends in the durability and final appearance of any stucco project. In cases where stucco finishes have not been satisfactory, the reason can usually be traced to the failure of masons to heed and follow such important directions.

FLASHING

Adequate flashing, using rust and corrosionproof metals, is also of great importance because it prevents water from getting through walls at various points. Ordinarily, flashing is required at tops and sides of projecting trim, under coping and sills, at intersections of walls and roofs, under built-in gutters, around all openings in roofs, at intersections of chimneys and roofs, and at any other points where moisture might gain entrance.

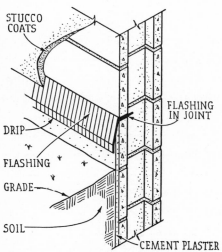

FIGURE 1. Flashing where no belt or soldier course is provided.

Figure 1 shows the proper application of flashing in a concrete-block wall at the point, near the grade line, where the stucco terminates. This is a plain block wall without a belt or soldier course at grade line. The flashing shown prevents water, which runs down the face of the wall on the stucco, from gaining entry through joints or from getting behind the stucco. Note that the flashing should be embedded in one of the mortar joints between the blocks.

Figure 2 shows how flashing should be used in connection with a concrete-block wall when a water table or belt course is built into the wall. When the flashing is properly embedded in a mortar joint, all water is directed away from joints and the back side of the stucco. Note that a drip is required to prevent water from gaining entry through the water-table joint.

Figure 3 shows how flashing should be used in connection with a metal window frame in a concrete-block wall.

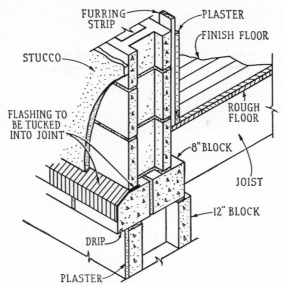

FIGURE 2. Detail of flashing with water table or belt course.

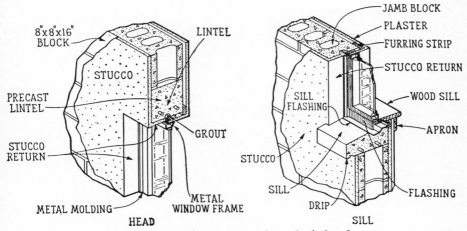

HEAD   SILL

FIGURE 3. Details of stucco around metal window frame.

Wherever stucco is to be applied over flashings or other surfaces which are unsatisfactory for direct application of it, metal reinforcement should be applied so as to provide a grip for the stucco. For example, where large areas of metal flashing are used, the stucco cannot stick to the metal unless a metal lath or other similar reinforcement is attached to the flashing prior to the application of the first coat of stucco.

TYPICAL DETAILS

The building codes in most cities and towns specify definite rules for stucco details. The following are typical:

Eastern Construction. In climates where cold or freezing weather occurs, stucco construction, used with frame walls, is generally somewhat similar to that shown in Figure 4.

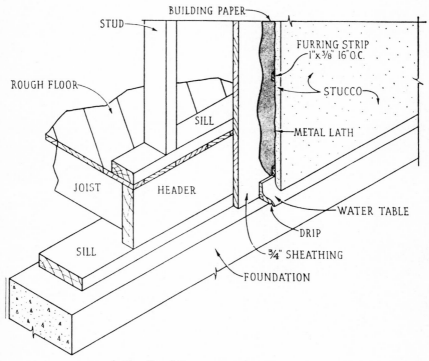

FIGURE 4. Details of Eastern-practice stucco construction.

Wood sheathing is applied to the studs. Over the sheathing, some form of waterproof building paper is attached. Furring strips, in either horizontal or vertical positions and spaced 16 inches apart on centers, are nailed over the building paper. Metal lath is applied to the furring strips and stucco to the metal lath.

This method of construction provides a wall which, because of the sheathing, building paper, and air space between the stucco and the sheathing, is warm and completely dampproof. Note the water table which directs water away from the top of the foundation. Metal flashing could be used in place of the wood water table.

Western Construction. In climates which are mostly warm and dry, stucco construction with frame walls is generally similar to that shown in Figure 5.

FIGURE 5 (*X*). This part of the illustration shows the ordinary 2 by 4 studs, spaced 16 inches on centers, which are generally used. The first step in stucco construction is to extend continuous wire, at about 8-inch intervals, all around a building, as indicated by the dashed lines *ah* and *jk*. The wire is applied as tightly as possible, Then, in order to make the wire even tighter, the length *ac* is pulled up to position *d* and there nailed. In like manner, the length *jm* is pulled up to position *n* and there nailed. This same procedure is followed for each wire all the way around a building. The wires are then able to resist pressure against them as stucco is being applied.

The purpose of the wires is to provide support for building paper and lath during the time the stucco is applied. As the stucco is being applied, the mechanics have to exert some pressure with their trowels. The wires are able to resist such pressure and tend to keep the building paper, metal lath, and stucco in proper position. Care should be taken to make sure that each wire is securely nailed to each 2 by 4 stud.

FIGURE 5 (*Y*). After all the horizontal wire is in place, a waterproof building paper is well lapped and nailed to the studs, using special nails which have large heads. The paper should be nailed, at intervals of about 8 inches, to each stud.

FIGURE 5 (*Z*). After the paper is in place, a special welded stucco wire should be nailed to the studs. Several varieties of the special wire required are available. This stucco wire, or, actually, reinforcement for the stucco, must be nailed to the studs using an approved type of furring nail. Such nails have double heads which, when driven, space the reinforcement about ¼ inch away from the building paper. There must be at least that much space between the reinforcement and the paper so that, when the stucco is applied, the reinforcement will be well embedded in it. The

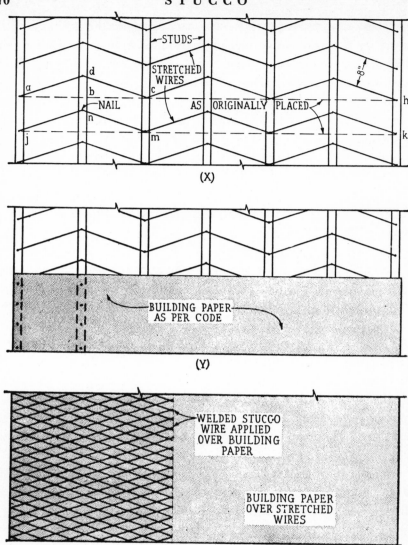

FIGURE 5. Details of Western-practice stucco construction.

reinforcement must be nailed to each stud at intervals no greater than 8 inches.

When the stucco hardens, it forms a thin but durable reinforced-concrete shell. If the wire, paper, and reinforcement have been properly constructed, the stucco will be able to resist wind pressure and earthquake shocks.

Stucco on Chimneys. Chimneys, because of the hot products of combustion which pass through them, may expand and contract from time to time. If stucco is applied to them without reinforcement, it might crack and even fall off as expansion and contraction take place. In order to avoid such possibilities, a metal reinforcement, such as shown in the Z part of Figure 5, should be securely attached to the chimney surfaces, above the roof line, before stucco is applied.

Wood Window Frames. When wood window (and door) frames are used in connection with frame walls, they constitute the head, jamb, and sills. The stucco should be applied flush with the outer edges of the frames, as indicated in Figure 7.

Metal Window Frames. When metal window frames are used in concrete-block walls, they are generally installed as shown in Figure 3. Note that there is a stucco return.

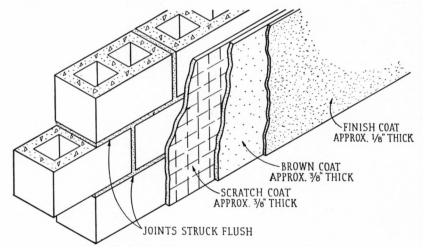

FINISH COAT
APPROX. 1/8" THICK

BROWN COAT
APPROX. 3/8" THICK

SCRATCH COAT
APPROX. 3/8" THICK

JOINTS STRUCK FLUSH

FIGURE 6. Application of stucco to concrete-block wall.

Stucco on Concrete Block. Figure 6 shows how stucco is applied to concrete-block walls. Note that joints between blocks should be carefully struck.

STUCCO-MIX PROPORTIONS

All coats of stucco should consist of one part cement and between three and five parts of damp, loose sand to which may be added hydrated lime not to exceed one-fourth the volume of the cement. Hydrated lime putty, slaked lime putty, or diatomaceous earth may also be added, for the sake of plasticity, in the same proportion as hydrated lime.

There is a great variety of stucco cements now available which can be mixed with sand on the job.

Finish Coat. When a white finish coat is desired, white cement and light-colored sand may be used. When colored finish coats are desired, factory-mixed stucco cement should be used.

MIXING

Care should be taken not to use cement which contains lumps. If the lumps cannot be pulverized, as explained in Chapter 4, Concrete, the cement should not be used.

The dry ingredients should be thoroughly mixed to a uniform color. Pure water (fit to drink) should then be added and the mixing continued until the mixture is in a uniformly plastic and easily workable state. When necessary, plasticity can be restored *without* adding any more water. This can be accomplished by remixing. In other words, work the mixture with a hoe or shovel.

THICKNESS OF STUCCO

For durable, strong stucco work, the total thickness should not be less than ⅞ inch. The recommended thickness of each coat is indicated in Figure 6.

APPLICATION OF STUCCO

If the first (scratch) coat is to be applied to concrete block, as shown in Figure 6, the surface of the block should be dampened as previously explained.

Scratch Coat. The scratch coat should be applied to a depth of about ⅜ inch. On frame walls, as shown in Figure 5, the scratch coat should cover all reinforcement. Application is by steel trowel. Inexperienced masons may find it necessary to use a screed to keep the surfaces of uniform thicknesses and true. Before this coat sets, it should be *scratched,* using a long wire brush made for that purpose. The scratches afford good bond for the second coat. This coat should be allowed to dry for at least 48 hours.

Brown Coat. Before this coat is applied, the surface of the scratch coat should be dampened. The application of this coat requires the use of screeds as shown in Figure 7. Prior to the time this coat is applied, various lengths of screeds should be available. Corners, such as *gq* and *hr* must have screeds such as shown in the *X* part of the illustration. The screed should extend ⅜ inch out from the wall.

Window frames, as shown in the *Y* part of the illustration, can serve as screeds. A long screed, as shown at *t,* can be stretched between the corner screed and the window frames and zigzagged back and forth to level and true up the stucco between the corner and the windows. Corner and other screeds, as shown at *a-a, c-c, e-e, b-b, d-d,* and *f-f,* can be lightly nailed in place. The screeds, such as at *t,* must be worked so that the surface of the brown coat is ⅛ inch below the edge of the window frame. This allows room for the final coat.

As the brown coat is applied, stretcher screeds, such as *g-g, h-h, o-o, p-p, g-g,* and *r-r,* can be used to straighten and true up the stucco to a depth of ⅜ inch.

After the stucco surfaces are straight and true, the screeds can be removed and the furrows they leave filled with stucco.

When the screeds have been removed, the surface of the brown coat can be worked, using a wood trowel or float, to make it more true. If necessary, a little water can be applied by means of a brush during the final troweling operations.

The surface of the brown coat should be left in the slightly rough condition caused by the wood trowel.

This coat should be allowed to dry for about 7 days before the final coat is applied.

Final Coat. This coat is applied in whatever color may be desired. It should not be more than ⅛ inch thick. It can be applied with a steel trowel but should be straightened and made true, using a wood trowel. A little water may be applied, if necessary, to help the smoothing procedure.

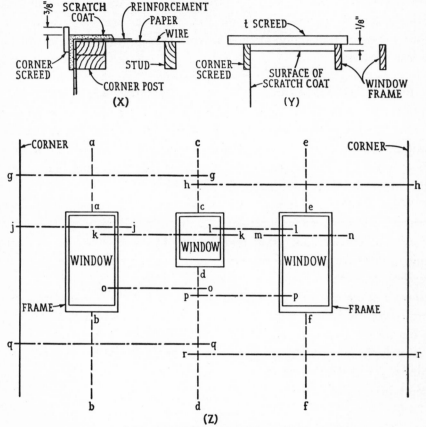

FIGURE 7. Use of screeds in applying stucco to a wall.

Various textures are possible in this final coat. The most common is made by using a wood trowel and by simply smoothing. Other textures can be made by swinging the wood trowel in both long and short arcs and not smoothing the slight ridges between the arcs. Still other finishes are possible by roughing the surface more than usual and then smoothing the high ridges with a steel trowel.

CHAPTER THIRTEEN

Insulation

There are thousands of wintertime situations in which people complain of feeling cold sensations and drafts, even though they are in practically airtight rooms which are kept at a temperature of 72°F by central heating plants. Generally, such situations bring about an increased and unnecessary consumption of fuel. During the summer, a reversed feeling of discomfort exists.

Farm animals are similar to people, so far as their comfort is concerned. If they feel comfortable, they produce, grow, and generally react favorably. On the other hand, if they are uncomfortable because of drafts, cold, or dampness, then decreased milk production, fewer eggs, and other unprofitable results occur.

Both houses and farm buildings can be comfortable, or they can be, as the old expressions have it, "cold as a barn" or "hot as an oven." The difference between comfort and discomfort is, to a large extent, a matter of planning and installing insulation wisely and well.

The purpose of this chapter is to explain briefly what insulation is, how it functions, and some of the important aspects of its use, especially as it is related to masonry walls, floors, and foundations. An explanation of the technical aspects of insulation, such as the calculation of heat losses and gains, theory of condensation, etc., is beyond the scope of this book, in which most of the explanations are concerned with manipulative procedures. Such aspects should be dealt with by architects and air-conditioning engineers. In unusual circumstances, not covered by this chapter, readers are urged to seek the advice of architects and engineers.

WHY DO WARM ROOMS OFTEN SEEM COLD?

In order to answer such a question, three simple factors must be understood. They have to do with laws of nature.

Conduction. When a warm substance comes in contact with a cold substance, the heat in the warm substance tends to pass from the warm substance to the cold substance. In other words, heat travels or flows from warmth to cold. This process is known as *conduction*. The term *substance* includes building materials and air.

Radiation. The surface of any object—building materials or people— will *radiate* heat to a nearby and colder surface. This is another application of the principle that heat flows from warmth to cold.

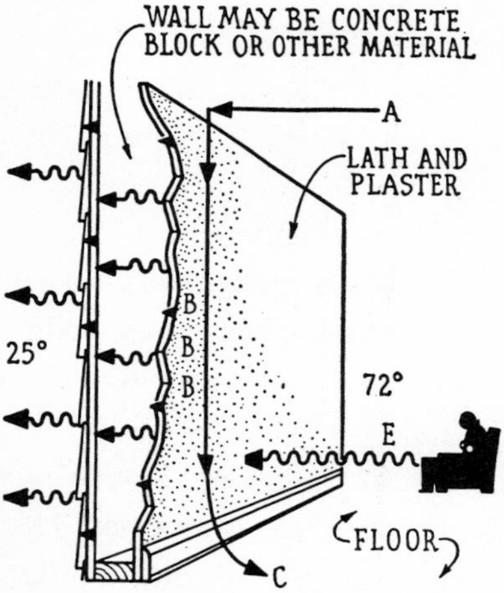

FIGURE 1. Why drafts are felt in a warm, closed room.

Convection. When a movement of air, as from one part of a room to another, is in process, heat can be transferred from one place to another. This transfer of heat is known as *convection*.

Example. Suppose that a person is seated at rest in a room having one exterior wall, such as indicated in Figure 1. Further suppose that the room temperature is 72°F and that the outside air has a temperature of about 25°.

It can properly be assumed that the building materials in the exterior wall are colder than the 72° air in the room. The wall is in contact with the room air, so by the principle of conduction the room air loses heat to the wall.

The wall is also in contact with the colder outside air. Thus, by the principles of *conduction* and *radiation* heat flows from the wall to the outside air.

As the room air, which is in contact with the wall, loses a considerable amount of its heat to the wall, it becomes colder and heavier and tends to drop down to the floor. As this happens, air from other parts of the room flows in to take the place of the air which dropped down to the floor. This process is illustrated in Figure 1. Air at position *B* becomes colder and drops down to the floor to position *C*. Warm air at position *A* flows in to take the place of the air which fell from position *B*. Thus, a movement or *draft* of air is produced, and heat flows by convection.

While all the foregoing is in process, the person seated at rest actually *radiates* body heat to the colder wall surface, as shown by the arrow at *E*.

Thus, the person can rightfully complain of feeling a sensation of cold and a draft. Heat from the room is being constantly lost to the outside air, the person radiates body heat away, and there is a movement, or draft, of air in the room.

HOW CAN ROOMS BE MADE MORE COMFORTABLE?

From the foregoing general explanations, it can be seen that the wall is the cause of all the trouble. It is easy to understand that, if the wall could be given some means of resisting the loss of heat to the outside air, a great deal of the trouble would be avoided. The heat loss from the room would be reduced, and drafts would not occur. Also, the interior surface of the wall would be warmer, and less body heat would be radiated to it by people.

Obviously, the remedy is to add some material to the wall which can resist the flow of heat and therefore avoid a great deal of heat loss. This can be accomplished by adding one of the various forms of insulation to the wall.

WHAT IS BUILDING INSULATION?

Briefly, building insulation is any material which has the ability to resist the flow of heat through it.

Some materials, such as iron and other dense and heavy, substances, have little or no resistance to the flow of heat through them. For example, if one end of an iron rod is put into a fire, the heat will travel to the other end of the rod within a short time.

Other materials, such as wood and similar less dense and lighter substances, have limited ability to resist the flow of heat through them. For example, if one end of a stick of wood is put into a fire, its other end, prior to actually burning, will not become too hot to handle.

Some materials, such as shiny aluminum foil, have the ability to reflect heat and thus prevent the heat from passing through them. For example, if two pieces of metal, one dull like iron and the other one shiny like aluminum foil, are placed where the rays of the sun can strike them, the iron will feel much hotter than the aluminum. This is because the shiny aluminum reflected the heat instead of allowing it to flow through it.

Insulation can thus be defined in two ways: first, as materials which resist the flow of heat because of their internal structures and, second, as materials which reflect heat away or resist its passage.

TYPES OF INSULATION

Insulation materials can be classified into the following general types, all of which are manufactured in various shapes and sizes:

1. Loose fill
2. Blanket
3. Bat
4. Rigid
5. Slab
6. Reflective

Loose Fill. This type of insulation is generally manufactured in loose form and sold in bags. It is poured, placed by hand, or blown into places where it can be confined to desired locations.

Blanket. Blanket insulation is flexible and usually manufactured in rolls. The material may be about 15 inches wide and 1 or more inches thick.

Bat. Bat insulation is similar to the blanket type except that the bats are usually about 15 inches wide, 15 inches long, and 1 or more inches thick.

Rigid. This type of insulation is sometimes known as structural insulation board. It is rigid, like boards, and is manufactured in various-sized sheets which are from ½ to 1 inch thick.

Slab. Slab insulation is manufactured in small rigid units 1 or more inches thick and ranging up to about 24 or 48 inches long, and sometimes longer.

Reflective. This type of insulation is manufactured, using aluminum foil, to create accordionlike sheets and other forms.

HOW INSULATION FUNCTIONS

As already indicated, insulation materials function in two general ways. First, they resist the flow of heat by means of their internal structures. Second, they reflect heat.

The first five types of insulation in the foregoing classification are so manufactured that their bulks actually contain billions of tiny dead-air cells or other dead-air spaces which resist the flow of heat by conduction, radiation, or convection. A dead-air space constitutes the best kind of insulation.

Reflective insulations simply reflect heat rather than allow it to pass through them.

Figure 2 shows three kinds of wall construction. The wall at *A* is composed of studs and sheathing. Wood has some ability to resist heat flow but not enough to be of much help in keeping the interiors of houses or other buildings warm. Therefore, such a wall only resists 5 out of 100 heat units. In other words, 95 out of every 100 units (a unit, for the purpose of this discussion, is simply an amount of heat) flow through the wall and are lost to the outside air.

The concrete-block wall or foundation at *B* has more resistance to heat flow through it, and only 63 units, out of the 100, are lost.

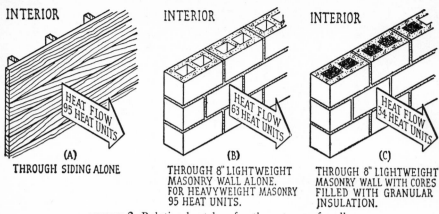

(A)
THROUGH SIDING ALONE

(B)
THROUGH 8" LIGHTWEIGHT MASONRY WALL ALONE. FOR HEAVYWEIGHT MASONRY 95 HEAT UNITS.

(C)
THROUGH 8" LIGHTWEIGHT MASONRY WALL WITH CORES FILLED WITH GRANULAR INSULATION.

FIGURE 2. Relative heat loss for three types of walls.

FIGURE 3. Heat loss in concrete-block wall or foundation is decreased by filling cores with granular insulation. (Courtesy of Portland Cement Association.)

The wall or foundation at *C* has its cores filled with a granular (loose-fill) type of insulation and thus loses only 34 out of every 100 heat units.

It should be noted that *lightweight* concrete block has much greater resistance to heat flow than ordinary *heavyweight* block.

Figure 3 shows how granular insulation is used to fill the cores of a block wall or foundation.

Core insulation can only be used to economical advantage when no reinforcement, in the form of rods or studs, is required in the walls or foundations.

HOW TO INSULATE CONCRETE-BLOCK AND PLACED-CONCRETE WALLS AND FOUNDATIONS

In addition to the use of granular insulation in the cores of plain block walls and foundations, there are several other methods which can be used with reinforced structures.

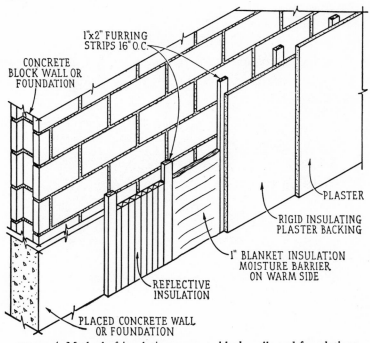

FIGURE 4. Method of insulating concrete-block walls and foundations.

Figure 4 shows what can be either a concrete-block or placed-concrete wall or foundation. The various types of insulation are all applied to the interior surface, and furring strips are required. Such strips can be attached to the block or concrete by means of special nails which are tempered and hard enough to be driven into the concrete without injury to the concrete or the nails.

Reflective Insulation. Several forms of this general type of aluminum insulation are available in sizes and shapes which fit between the furring strips. The insulation generally has nailing strips or flanges which can be fastened to the furring.

Blanket Insulation. Several forms of this general type of flexible insulation are available which also fit between, and are flush with, the furring strips. Nailing flanges are generally provided.

Rigid Insulation. Any of the several forms of this general type can be nailed to furring strips and used as surfacing for walls or foundations, or the insulation can be used as a plaster backing. Decorative forms are also available.

Any one or a combination of the foregoing insulation methods can be used to advantage in giving walls and foundations much greater ability to resist the flow of heat.

HOW TO INSULATE
CONCRETE SLAB FLOORS ON GROUND

In the Chapter 7, Concrete Floors, mention was made of fills as a means of helping to insulate slabs on ground. That method is effective and serves a good purpose.

Figure 5 shows two other effective ways of insulating slabs which might be parts of radiant-heating panels.

Insulating Concrete. This type of concrete is made using lightweight and porous aggregate. There are several available commercial products, such as Perlite. The resistance to the flow of heat is definitely appreciable and worth the expense involved. Such concrete is mixed in the usual manner, except that much more water is necessary. For example, a typical Perlite mix is made using 1 bag of cement, 4 cubic feet of Perlite, and 9 gallons of water. This kind of concrete is placed in the normal way but should not be used where strength is required.

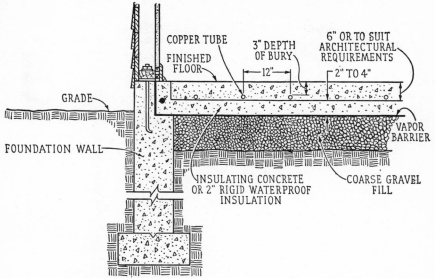

FIGURE 5. Use of insulating concrete to keep concrete slabs on ground warmer and less damp.

Rigid Waterproof Insulation. Two inches of this type of insulation can be used with effective results. However, unless the finished grade and fill are especially well packed and firm, the regular concrete, which constitutes the floor, is apt to crack if heavy weight is applied to it at a particular spot or area.

HOW TO INSULATE
PLACED-CONCRETE BASEMENT FLOORS

Figure 6 shows how a placed-concrete basement floor can be insulated while at the same time providing a floor surface which will feel almost as soft as a carpet.

Large-sized sheets of 1-inch rigid insulation should be cut and fitted so as to cover the concrete surface completely. Then, remove the insulation and apply hot bitumen over the floor. Re-lay the insulation. It will adhere to the bitumen. The bitumen also creates a moisture barrier between the concrete and the insulation.

Next, cut and fit ¼-inch hardboard sheets so that they completely

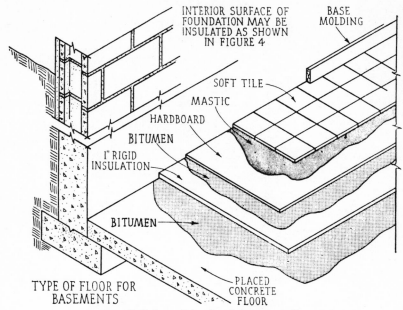

INTERIOR SURFACE OF
FOUNDATION MAY BE
INSULATED AS SHOWN
IN FIGURE 4

BASE
MOLDING

SOFT TILE

MASTIC

HARDBOARD

BITUMEN

1" RIGID
INSULATION

BITUMEN

TYPE OF FLOOR FOR
BASEMENTS

PLACED
CONCRETE
FLOOR

FIGURE 6. Soft and insulated basement floor.

cover the insulation. Cement down with more hot bitumen. The purpose of the hardboard is to distribute weights, such as from chair and table legs, over a greater area and thus prevent dents in the relatively soft insulation.

Finally, any kind of soft tile can be cemented to the hardboard and a base molding applied around the wall. The resulting floor is dampproof, quiet, soft, and an excellent addition to any basement living quarters or recreation rooms.

HOW TO INSULATE CRAWL SPACES

Figure 7 shows a section view of a typical crawl space and several methods of insulation to prevent cold, damp floors above such spaces.

Reflective Insulation. This type of insulation can be applied between the joists as indicated. It should be fastened to the joists on both sides by its nailing flanges.

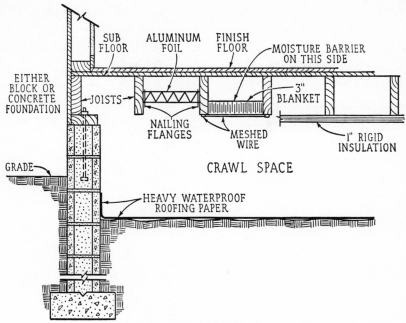

FIGURE 7. Methods of insulating crawl spaces.

Blanket Insulation. This type of insulation should be at least 3 inches thick and should have a moisture barrier on one side.

Fasten strips of galvanized wire or other weather-resisting material taut across the underside of the joists. The insulation should be shoved between the joists. The moisture barrier side should be *up*—that is, facing the floor.

Rigid Insulation. One-inch rigid insulation can be securely nailed to the bottom edges of the joists. It should have a moisture barrier on its upper side the same as the blanket insulation.

HOW TO INSULATE FARM BUILDINGS

Buildings which house livestock and poultry should generally be kept at a temperature of about 40°F during the winter months. The interior humidity conditions should not be in excess of about 80 per cent.

Insulated walls reduce both heat loss and the condensation which causes humidity. This means that more of the heat given off by livestock can be saved to warm the air brought in by ventilation. In turn, more fresh air can be brought in to replace stale high-humidity air without cooling buildings too much. Insulated walls actually make buildings drier.

As previously explained, lightweight concrete block and granular insulation can be used to make walls more resistant to heat loss. Cavity walls also provide appreciable resistance to heat loss. However, the roofs and ceilings of barns should be insulated too.

Table 1 shows the locations and amounts of insulation to use for barns located in various zones of the country. Figure 8 shows the four zones.

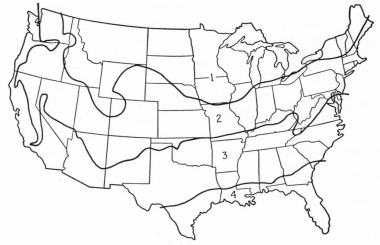

FIGURE 8. Climate zones.

In Zone 1, the walls of buildings should, if possible, be built with lightweight blocks and their cores filled with granular insulation. If lightweight blocks are not available, cavity walls can be constructed.

In Zone 2, lightweight blocks without the granular fill can be used.

In Zone 3, and also Zone 4, regular concrete block can be used.

HOW TO INSULATE
PLACED-CONCRETE ROOFS

Concrete roofs can be effectively insulated by applying rigid slab insulation over a bitumen mopping. Generally, such insulation is ap-

TABLE 1. Insulated farm-building construction

BUILDING / WALL	DAIRY BARN LARGE	DAIRY BARN SMALL	MILK HOUSE	POULTRY HOUSE	HOG HOUSE	BUILDING / CEILING
CONCRETE BLOCK	3	3	3	3	3	
CAVITY WALL 2-4" WALLS 2" AIR SPACE	2	3	2	2	2	
LIGHT WEIGHT BLOCK	2	3	2	2	2	
LIGHT WEIGHT BLOCK INSULATION IN CORES	1	1	1	1	1	
CAVITY WALL 2-4" WALLS 2" INSULATION	1	1	1	1	1	

NUMBERS REPRESENT USDA CLIMATIC ZONES

plied 2 to 4 inches thick. Built-up roofing can be applied over the insulation.

HOW MUCH INSULATION TO USE

The amount of insulation to use is a problem which should be given careful thought because of the law of diminishing returns. This law, in terms of insulation, means that the cost savings resulting from prevention of heat loss with insulation is not proportional to the thickness of the insulation. The use of *two* types of insulation, or two layers, instead of one, further reduces heat loss during the winter or heat gain during the summer. However, the savings is not double just because double the amount of insulation is used. In other words, the use of *both* blanket and rigid insulation, as shown in Figure 4, does not save twice as much money in heating or cooling costs as the use of either type.

Under ordinary circumstances, the use of lightweight block and any *one* of the insulations shown in Figure 4 constitutes the most economical use of insulation.

HOW TO PREVENT CONDENSATION ON BASEMENT WALLS

The air in all parts of houses and other buildings or out of doors contains moisture in the form of an invisible vapor. The warmer the air, the more water vapor it can hold. In other words, as air becomes warmer, it can hold increasing amounts of water vapor. Then, when air is cooled, it cannot hold as much vapor and condensation occurs.

Example. During the summer months, a great deal of moisture is entrained (held) by warm air. Such air is called *humid* and often causes discomfort. If such air comes in contact with a cold surface, such as a glass containing ice-cold water, the air in the immediate vicinity of the glass is chilled. As it chills, it is no longer able to hold as much water vapor, with the result that the outside surface of the glass becomes running wet. Some of the water vapor in the air is actually deposited on the glass.

Foundations, such as basement walls, are in contact with the soil at depths where the soil is cool. Therefore, the foundations stay cool. When humid basement air comes in contact with the cool surfaces of the foundations, its temperature is lowered, with the result that some of its water vapor is deposited on the surface of the foundations. In some instances, the surfaces of foundations become running wet. This can be avoided by using rigid insulation, as shown in Figure 4. The insulation prevents warm and humid air from coming in contact with the cool surface of the foundations.

SOUND CONTROL

As explained in Chapter 8, Concrete Block, the surfaces of concrete-block walls, for example, do absorb appreciable sound and thus help to make rooms more quiet. The use of lightweight porous block, such as shown in Figure 3, helps to absorb considerable sound.

Even more soundproofing can be accomplished by the use of open-textured and soft rigid panels applied as shown in Figure 4. Several types of special insulation are available which absorb sound. Any one of them can be applied to block or placed-concrete walls and foundations. They can also be cemented to ceilings under placed-concrete floors.

Acoustical Plaster. Various insulating materials are available which can be substituted for sand in making plaster. Such materials contain a high percentage of dead air-cells which make them light in weight and able to absorb sound effectively.

CHAPTER FOURTEEN

Illustrative Example

In the foregoing chapters, commonly encountered aspects of *placed-concrete* and *concrete-block* construction, as generally applied to the planning and erection of houses and other small buildings, are explained and illustrated. The presentation shows the application of proved principles, the use of proper planning, and suggested manipulative procedures, all of which are intended as a means of helping young mechanics and other readers to accomplish satisfactory, economical, and durable masonry work.

The purpose of this chapter is to show how such aspects are involved in the planning and erection of a typical building for which the foundation, floor, and walls are specified in terms of placed concrete and concrete block.

PLAN OF CHAPTER

Figures 1 through 6 show the working drawings (which have to do with masonry work) for an apartment building which is to have a living room, two bedrooms, and two bathrooms. This chapter explains and illustrates the planning and erection aspects, so far as placed-concrete and concrete-block structural details are concerned.

In keeping with the importance and desirability of reinforcement, the working drawings specify what is probably the maximum amount of steel likely to be required by any building code. However, the same building could be erected with much less, or no reinforcement at all, except in the lintels which would then be necessary over all window and door openings.

With or without reinforcement, the procedures for placing concrete and for laying up block are the same.

The working drawings obviously deal with a building to be erected in a mild climate where interior furring and plastering of walls and ceilings are not necessary if structural economy is of concern. Also, the windows are large and typical of a climate where severe cold and winds are not likely. However, the same general design could be applied to other climates with but a few planning changes. Thus, the building serves the purpose of this illustrative example in which procedures usable in any climate or region of the country are described.

All through this chapter, the erection procedures are presented in terms of minimum special equipment or tools. For example, it is assumed that surveying instruments are not available. However, where such equipment is available, it can be substituted for the practical methods herein.

As in preceding chapters, all the suggested manipulative procedures are based upon practices observed in all parts of the country but are not offered in the sense of being the only acceptable methods.

The working drawings shown in Figures 1 through 6 are representative of such drawings as are required by cities and towns as a means of obtaining permits.

PLANNING

Ordinarily, masons are not concerned with the architectural aspects indicated in working drawings. In other words, room sizes, room shapes, window sizes and locations, general appearance, etc., are established directions created by architects which are to be used by masons as a guide. However, masons should study working drawings to learn what is required of them and to check such items as may be important to their work, as explained in the following:

Plot Plan. The plot plan, as shown in Figure 1, shows the exact position of the proposed annex building and is shaded to set it apart from the existing building. This plan shows that the south side of the annex is to be 8′ 0″ from the existing building, 5′ 0″ from the west property line, and 8′ 0″ from the north property line. It also indicates that the annex should adjoin the existing washhouse. This plan is therefore used to lay out the

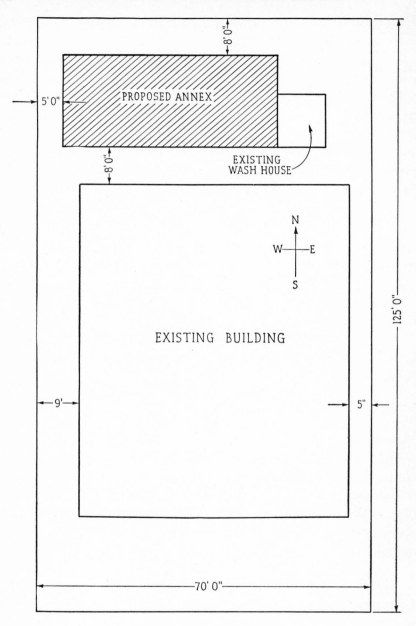

PLOT PLAN
109 VISTA DEL MAR
1" = 20' 0"

FIGURE 1. Plot plan.

position of the required foundations. No contour lines appear, so it can be assumed that the lot is flat and level.

Footing Plan. This plan, as shown in Figure 2, gives the over-all dimensions of the annex, the required foundation width, and several notes. For example, a 4-inch concrete floor is required. The concrete used should be a 1: 2½: 3½ mix. The floor is to be troweled smooth to receive asphalt tile. The notes indicate that lightweight grade N-I blocks must be used and that they must conform to certain specifications.

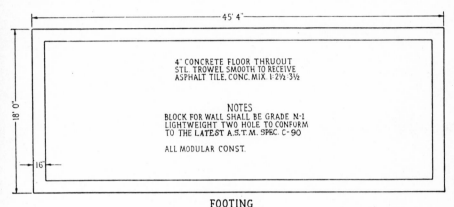

FOOTING
¼" = 1' 0"

FIGURE 2. Footing plan.

Floor Plan (see Figure 3). So far as masons are concerned, the floor plan is used only to indicate the size and locations of all windows and doors, the positions and spacing of partition anchors, and places where holes will have to be cut in blocks for electrical outlets and switches.

The window and door sizes are shown. For example, window *A* for bedroom 2 is to be 6' 0" by 4' 0". This means 6' 0" wide and 4' 0" high. The doors are to be 3' 0" plus 2 inches for each jamb or 3' 4" wide. Door heights are given in the elevation views.

Partition anchors are to be placed on the center lines of all partitions and at 4' 0" intervals.

Electrical outlets will be spotted by electricians who place conduit prior to the time the floor slab is placed.

All plumbing lines will be placed by plumbers before the floor is placed.

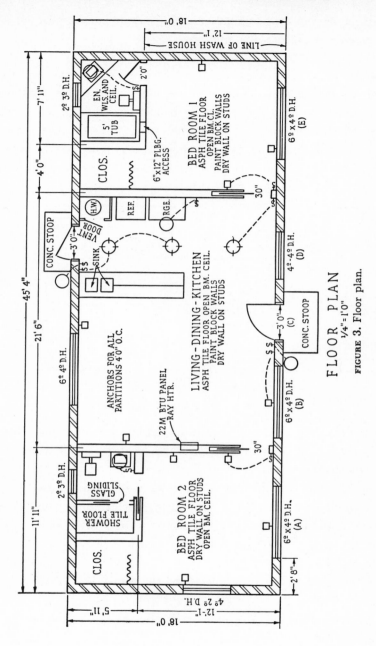

FLOOR PLAN
1/4" = 1' 0"

FIGURE 3. Floor plan.

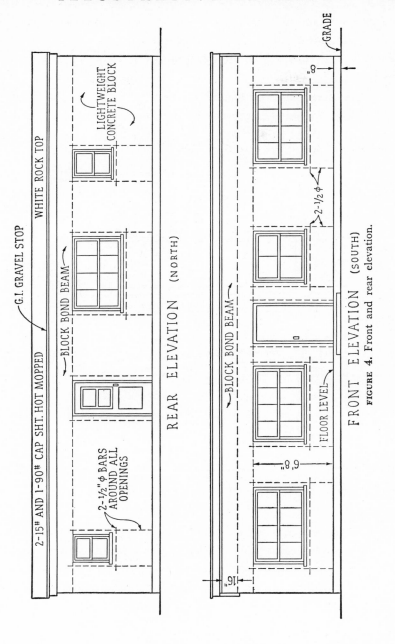

FIGURE 4. Front and rear elevation.

Front and Rear Elevations (see Figure 4). These views show a considerable amount of information necessary to masons.

WINDOW HEIGHTS. All window heights are to be 6′ 8″ above the floor level.

DOOR HEIGHTS. All door heights are to be 6′ 8″ above the floor level.

FLOOR ELEVATION. The surface of the floor is to be 8 inches above grade.

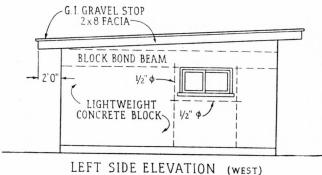

LEFT SIDE ELEVATION (WEST)

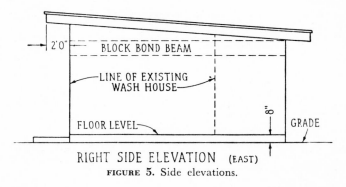

RIGHT SIDE ELEVATION (EAST)

FIGURE 5. Side elevations.

BOND BEAM. There is to be a block bond beam which will serve to tie the walls together and as a lintel over all window and door openings.

REINFORCEMENT. Two ½-inch round steel bars are to be used on both sides of all openings and under the window sills. Vertical reinforcement is to extend from the footing up to and into the bond beam.

Side Elevations (see Figure 5). So far as masons are concerned, these two views show about the same information as the front and rear elevations.

Section (see Figure 6). This view, like the front and rear elevations views, shows a considerable amount of information necessary to masons.

FOUNDATION. The foundation must be made of placed concrete, be 22 inches high and 16 inches wide, and contain three ½-inch steel bars which are to be continuous. The top surface of the foundation is to be at the same natural grade as the ground.

FLOOR SLAB. The surface of the floor slab must be at the same elevation as the top of the first course of the wall block. A floor thickness of 4 inches is specified.

FILL. There is to be a 4-inch sand fill under the floor slab. NOTE: This deviates from suggestions given in a previous chapter but is satisfactory in a dry climate.

WALL HEIGHTS. The front wall is to be 10′ 0″ high, including the block bond beam and the rear wall 8′ 8″ high, including the bond beam.

DOWELS. One-half-inch round steel dowels are to be spaced at 24-inch intervals in all walls. They must extend 24 inches above the foundations and also extend down into it so as to loop the lower rods.

STUDS. One-half-inch round steel studs, extending from the top of the foundation up to and into the bond beam, must be placed at 24-inch intervals in the block walls.

BOND-BEAM STEEL. The bond beam is to have four steel rods in it.

ANCHOR BOLTS. Anchor bolts are required at the top of the wall as a means of anchoring the 2 by 8 plate. The bolts must be ¾ inch in diameter and be spaced at 48-inch intervals.

WALLS. The walls must be 8 inches thick, and their exterior faces should be flush with the exterior face of the foundation.

DIMENSION CHECKS

It is assumed that modular block will be used. Therefore, several dimensions should be checked before the foundations are laid out.

Over-all Dimensions. The over-all dimensions are to be 45′ 4″ and 18′ 0″. The exterior faces of the walls must be flush with the exterior

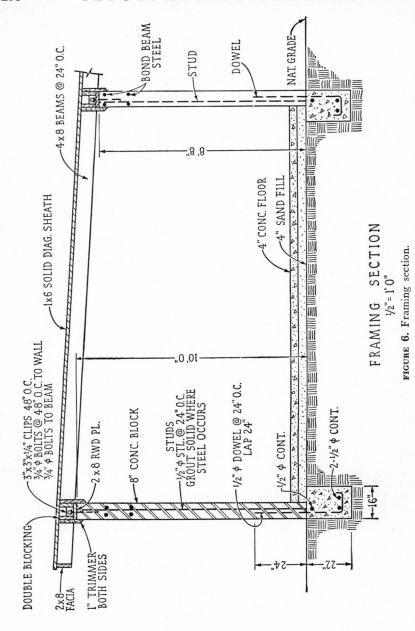

FRAMING SECTION
½" = 1'0"

FIGURE 6. Framing section.

face of the foundation. Thus, the exterior dimensions for the walls are also 45′ 4″ and 18′ 0″.

With a ⅜-inch mortar joint, each 15⅝-inch-long block constitutes a length of 16 inches. The 45′ 4″ dimension equals 544 inches which, divided by 16, equals exactly thirty-four of the 16-inch stretchers. Thus, that dimension is modular, and blocks will not have to be cut.

The 18′ 0″ dimension equals exactly 13½ stretchers, so it is accurate according to modular dimensions, too.

Height Dimensions. The section view in Figure 6 shows wall heights of 10′ 0″ and 8′ 8″. The 10′ 0″ dimension equals 120 inches, which, divided by 8 inches (the height of one 7⅝-inch course plus a ⅜-inch joint), gives exactly fifteen courses. The 8′ 8″ dimension equals exactly thirteen courses. Thus, both heights are modular. The difference between the two heights can be accounted for in terms of the roof pitch.

The front elevation view, as shown in Figure 4, contains a 6′ 8″ dimension which locates the window and door heights as well as the bottom of the bond beam. That dimension equals exactly 80 inches, or ten courses. It is therefore modular.

The window shown at *A* is 4′ 0″ high. That is also a modular dimension which contains exactly six courses. All other window and door heights check accurately.

Further help in visualizing the modular aspects of the dimensions is presented in Figures 9, 10, and 15.

If any such important dimensions do not check out accurately from the modular standpoint, the architect should be consulted.

FOUNDATION LAYOUT

Once the plans have been studied and all dimensions found to be modular, the foundation can be laid out following the simple procedures outlined in Chapter 5, Concrete Footings. When string has been stretched and all corners located, according to the dimensions shown in Figure 1, a check should be made to make sure the layout is exactly 8′ 0″ from the existing building, 8′ 0″ from the north property line, and 5′ 0″ from the west property line.

EXCAVATION AND FORMS

Assuming that the soil is firm, it can be used in place of complete forms, as shown in the Y part of Figure 7. The trench sides, shown at d and $e,$ can be used to hold the concrete in place as it hardens. However, good practice dictates the use of 2 by 8 forms for the upper part of the foundations. The forms assure a flat and perfectly level surface at the top of the foundation. They also provide square and true top corners which are used to advantage when laying the first course of blocks. The excavation and form erection can be carried out as a joint project.

First, a trench which is about 8 inches deep and 30 inches wide may be excavated under the layout strings, somewhat as shown in Figure 11 of Chapter 5, Concrete Footings. The 30-inch width allows space, as shown at a and $b,$ in the Y part of Figure 7, in which to work. The bottom of the trench should be fairly level.

The 2 by 8 forms may also be erected as explained in Chapter 5, Concrete Footings, and as shown in the X part of Figure 7. Stakes should be driven at about 4' 0" intervals along both sides of the forms.

After the forms are completed, they should be carefully checked to make sure that each side is perfectly level and that both sides are at the same level. This can be accomplished by the use of a long 2 by 4 and a level. Place the 2 by 4 and level in various positions across the forms, as indicated by the dashed lines in the W part of the illustration. It is important that all sections of the forms be absolutely level and that both sides be at the same level. Otherwise, there is bound to be serious trouble in laying up the block wall and in making the floor slab level.

Once the forms are complete, the balance of the excavating can be done, using care to keep the sides d and e as vertical as possible and not to disturb the stakes. If some of the soil between the stakes and the edge of the trench falls out, it can be removed and the concrete placed without difficulty. The bottom of the trench should be 22 inches below the top of the forms, consist of undisturbed soil, and be level. The level can be checked by the use of a long 2 by 4 and a level. Place the 2 by 4 and level along the bottom and excavate as required to assure a level bottom.

If, for example, the soil is composed of dry sand, it will not stand firm, and forms then have to be erected to the full depth of the foundation.

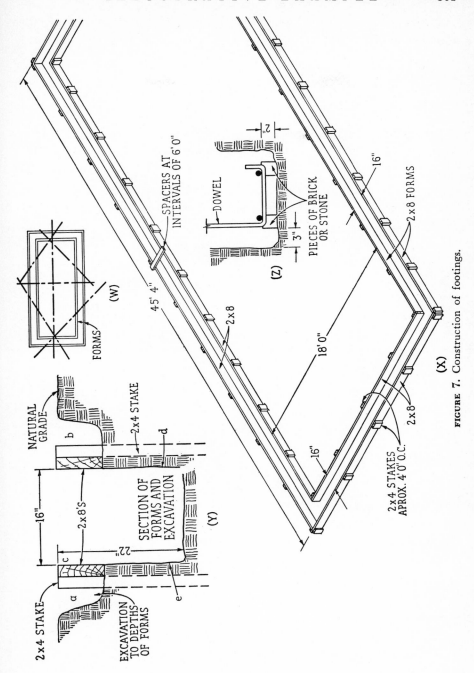

FIGURE 7. Construction of footings.

SPACERS AT INTERVALS OF 6'0"

DOWEL

2"

3"

PIECES OF BRICK OR STONE

(Z)

16"

2x8 FORMS

FORMS

(W)

45' 4"

2x8

18' 0"

16"

2x8

2x4 STAKES APROX. 4'0' O.C.

(X)

NATURAL GRADE

2x4 STAKE

16"

2x8'S

22"

SECTION OF FORMS AND EXCAVATION

(Y)

2x4 STAKE

EXCAVATION TO DEPTHS OF FORMS

a

b

c

d

e

Such forms are similar to those illustrated and explained in Chapter 6, Concrete Foundations.

PLACING REINFORCEMENT

Before the concrete for the foundation can be placed, its reinforcement and the dowels, as specified in Figure 6, must be placed and held in the proper positions.

Bottom Rods. The two bottom rods can be held about 2 inches above the bottom of the trench by the use of pieces of old brick or stones, as shown in the Z part of Figure 7 and in Figure 8.

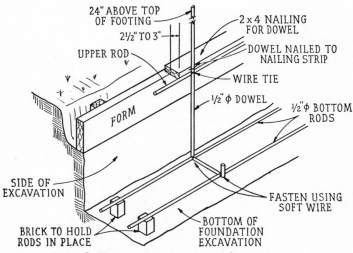

FIGURE 8. How to place dowels and foundation steel.

Dowels. Figure 8 shows a cutaway view of the trench and one of the 2 by 8 forms. Note that a piece of 2 by 4 lumber (nailing for dowel) is nailed to the top of the form. The dowels can be held upright by nailing them to that horizontal piece. The lower ends of the dowels should be wired to the bottom rods as indicated.

Upper Rod. The upper rod may be wired to the dowels, as shown in Figure 8.

If reinforcement is placed in the foregoing manner, it will stay in proper position during the time the concrete is being placed and as it hardens.

PLANNING WINDOW AND DOOR LOCATIONS

From previous checking, it was established that all dimensions are modular. However, it is a good idea for inexperienced masons actually to draw scaled sketches when planning window and door locations.

Figure 9 shows the full length of the front wall of the building indicated in Figures 1 through 6. Note that the 45′ 4″ length includes exactly 34 full stretchers. The opening marked *A*, for example, is the same as the opening marked *A* in Figure 3.

Figure 3 specifies that the tops of the windows and of the doors must be 6′ 8″ (exactly ten courses) above the floor level, that the windows must be 4′ 0″ high, and that the door must be 6′ 8″ high. The window heights equal exactly six courses and the door height exactly ten courses. Thus, there must be four courses between the sills of the windows and the floor level and no courses between the sill of the door and the floor level.

Figure 3 also specifies that the left jamb of window *A* must be 2′ 8″ from the corner of the wall. That distance equals 32 inches, or exactly the length of two full stretchers. Thus, as sketched to scale in Figure 9, the location of window *A* is established and checked.

By scaling, it can be seen that the left jamb of window *B* is 4′ 0″ from the right jamb of window *A*. That distance equals 48 inches which exactly equals the length of three stretchers. Thus, as sketched to scale in Figure 9, the location of window *B* is established and checked.

All other openings can be established and checked in the front and all other walls in the same manner.

PLANNING STUDS

Figure 6 specifies that studs shall be spaced at 24-inch intervals in the walls and from the top of the foundation up to and into the bond beam.

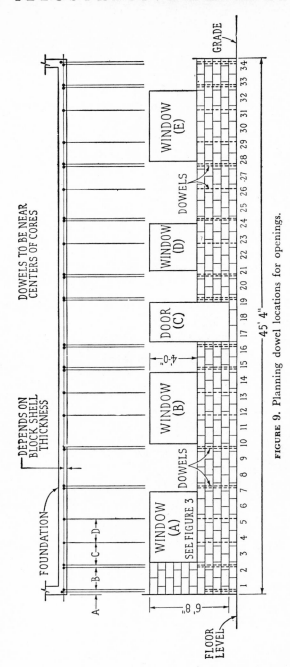

FIGURE 9. Planning dowel locations for openings.

Figures 4 and 5 specify that double bars be used on both sides of all openings. This makes for complicated planning unless a scaled sketch is used.

Figure 10 shows the same wall and openings as explained for Figure 9. The dotted lines, on both sides of the window and door openings, represent the double bars. The upper part of the illustration shows in which stretchers the bars must be.

Because the openings are so close together, the double bars practically serve the purpose of the specifications shown in Figures 3 through 5. With the exception of the space between windows D and E, no other spaces between openings require studs. One could be placed in the space between windows A and B, but that distance is not great enough to merit the additional steel.

Studs and double bars around all other openings are planned in the same manner.

PLANNING DOWELS

According to the specifications shown in Figure 6, dowels are required at 24-inch intervals in the walls. They must also lap the studs.

Because of the window and door locations and their required double bars, dowels can seldom be spaced exactly 24 inches apart. First, they must be spaced to lap the studs and double bars on each side of the openings. When that specification has been satisfied, only the parts of the walls under the windows require additional dowels. The dashed lines in Figure 9 represent the required dowels for the front wall. NOTE: The illustration shows double dowels to match the double rods on the sides of the openings. This practice is not absolutely necessary.

The upper part of Figure 9 shows the dowel locations in the foundation. Such locations must be known before the dowels can be wired into position, as explained for Figure 8, prior to the placing of concrete. For example, the dimensions A, B, C, and D, as shown in the upper part of Figure 9, must be known.

Inexperienced masons can easily and accurately determine such dimensions by placing a line of dry block along the ground in a straight line, so

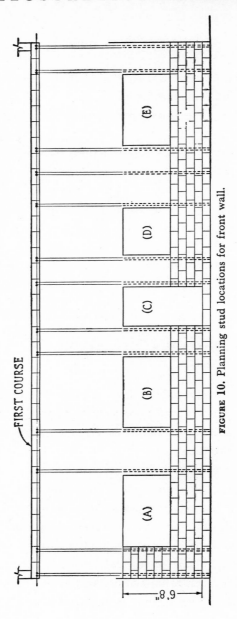

FIGURE 10. Planning stud locations for front wall.

that there are ⅜-inch spaces between blocks, something like the suggestion shown in Figure 11.

The blocks shown in the X part of the illustration correspond to the first six blocks of the first course shown in Figure 9. The A, B, C, and D dimensions shown in the X part of Figure 11, correspond with the like dimensions shown in the upper part of Figure 9.

From Figure 9 it can be seen that dowels appear in blocks 1, 2, 4, and 5. These locations are dimensioned, using like dimensions in the X part of Figure 11. Such distances can be actually measured along the row of

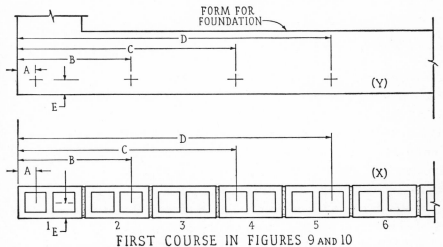

FIGURE 11. How to measure and lay out locations for dowels before placing them in foundation forms.

dry blocks and the resulting distances transferred to the trench, as shown in the Y part of Figure 11. Then, all dowels can be properly located and wired into positions.

PLACING FOOTING CONCRETE

The concrete should be placed in layers, as directed in Chapter 6, Concrete Foundations, starting at one corner and proceeding all the way around the forms. The first layer must be carefully spaded to make sure that the concrete completely surrounds the bottom bars. However, care must be taken to avoid moving the bars from their correct positions.

Succeeding layers should be placed, in a continuous operation, until the forms are completely filled. As the concrete starts to harden, the surface can be troweled with a wood float or trowel to make it smooth and true.

FORM REMOVAL

During warm weather the forms can generally be removed after 2 or 3 days. During cold weather, they should be left in place for at least 1 week.

PLACING FIRST COURSE OF BLOCK

Figure 6 indicates that the surface of the floor must be 8 inches above the top of the foundation. In other words, one course of block has to be laid before the concrete floor can be placed. Even though only one course of block is to be laid up prior to the placing of the floor, leads must be laid at all corners first, as shown in Figure 12. Laying up the leads should be done according to the same procedure explained in Chapter 10, Concrete-block Walls, Foundations, and Pilasters.

When all four leads are in place, each of them should be checked for level as indicated by the dashed lines x-x, y-y, and w-w. In addition, the level should be checked by placing a long 2 by 4 on edge, as indicated by the dashed line z-z. The long 2 by 4 should also be placed at other locations to make sure that all parts of the first course are at exactly the same elevation. A long measuring tape should also be used to check the 18′ 0″ and 45′ 4″ dimensions after the leads are laid up. If such dimensions do not check accurately, the leads should be removed and relaid. Keeping all dimensions exact is another item of great importance.

PLACING THE SAND FILL

When confined, as by foundations, sand constitutes a satisfactory fill material, especially in climates which are apt to be dry. It also forms a good base for a concrete slab and can be spread to an even level with ease.

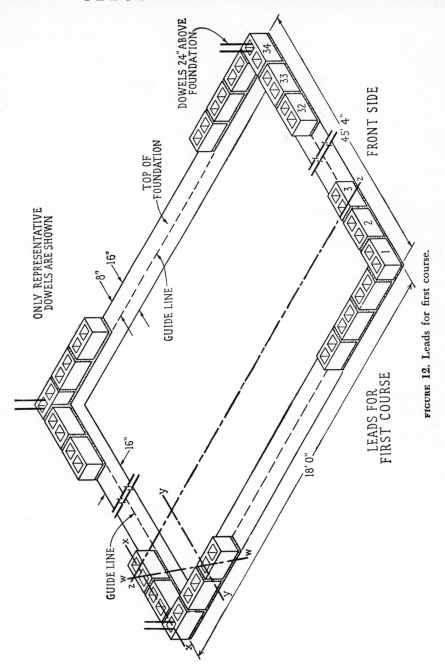

FIGURE 12. Leads for first course.

As indicated in Figure 6, a sand fill of 4 inches is required. The area of the fill region is roughly 45 by 18, or 810 square feet. Multiplying the area by 4 inches (⅓ foot) gives 270 cubic feet, or 10 cubic yards of sand required.

The sand can be dumped into the fill region and then spread with a shovel until it is at the approximate depth and smoothness desired. To obtain accurate depth and a true surface, a movable screed, such as shown in Figure 13, should be used. It is advisable to wet the sand before final spreading and smoothing by means of the screed, because, when wet, sand will compact and smooth to better advantage. The screed should be pushed along the foundations, back and forth, until the fill region is smooth and level.

If desired, a staked-up screed can be set along the center line of the fill region. Then a shorter movable screed can be handled by one man.

The sand should be well compacted along the interior surfaces of the foundation so that the concrete slab will have firm support.

PLACING CONCRETE FLOOR SLAB

In order to screed the concrete accurately and easily, a staked-up screed should be erected along the center line of the floor, as shown in Figure 14. The screed should be carefully checked to make sure it is level and at the same elevation as the tops of the first block course. The movable screed and a level can be used at several locations for checking purposes.

An hour or two before the concrete is placed, the sand fill should be thoroughly wetted, using a hose having a spray. The spray should be light so that no sand is displaced. The wetting is necessary as a means of preventing the sand from sucking too much water out of the newly placed concrete.

Concrete should be placed on only one side of the staked-up screed at a time, as explained in Chapter 7, Concrete Floors. When the first section is properly screeded, the second section can be placed. After both sections have been screeded and the concrete has been raked, the staked-up screed can be removed and its furrow filled with concrete.

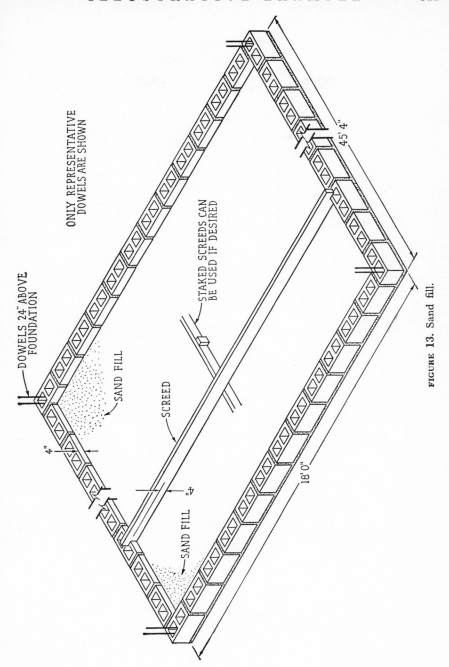

ONLY REPRESENTATIVE
DOWELS ARE SHOWN

DOWELS 24" ABOVE
FOUNDATION

STAKED SCREEDS CAN
BE USED IF DESIRED

SAND FILL

SCREED

SAND FILL

4"

4"

45' 4"

18' 0"

FIGURE 13. Sand fill.

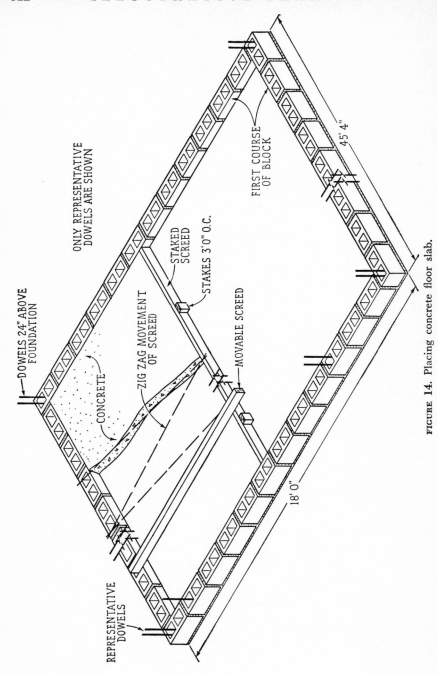

FIGURE 14. Placing concrete floor slab.

An area of this size requires the efforts of two men for wood and steel troweling. Such work is carried out as explained in Chapter 7, Concrete Floors. Power trowels are available, but the work can be done quickly enough by two or three men.

There may be some question as to why steel reinforcement was not specified for the floor. Actually, there is no need for it because the sand fill is confined and there is no possibility of its failing to provide good support for the floor.

Care should be taken to cure such a floor properly.

LAYING UP THE BLOCK WALL

As shown in Figures 9 and 10, the window openings are only four courses above the floor level. Thus, it is practical to lay up four-course leads at the four corners of the building and then lay up the intervening block as explained in Chapter 10, Concrete-block Walls, Foundations, and Pilasters.

Figure 4 specifies that reinforcement must be put in the walls under all window openings. Such reinforcement should be in the joint marked R in Figure 15 and may consist of the type shown in Figure 26 in Chapter 9, Typical Concrete-block Details.

When the four courses (five courses counting the first course) have been laid up and the mortar has hardened, the precast window sills can be placed. Full mortar beds should be spread and the sills carefully checked to make sure they are plumb in all directions. Two days should be allowed for the sill mortar to harden.

As the third or fourth course is laid up, the face of the wall should be marked with chalk to indicate the cores which contain the dowels and where studs will be placed just before the concrete is poured for the bond beam.

The window and door frames, as shown in Figure 15, may then be set and braced. To start with, lightly tack braces in both sides of all frames. Then use a level, as shown by the dashed lines *x-x, y-y,* and *z-z,* to plumb each frame. When the frames are plumb, the braces can be nailed tight. After nailing, the same plumb checks should again be made.

Along the tops of the heads, drive two or three large-headed screws well into the wood. When the bond beam is laid up, these screws will be embedded in the mortar and prevent any possibility of the wood heads' pulling away from the lower side of the beam if they should warp or shrink.

The blocks between frames should be laid up using the guide cord mentioned in previous chapters. The use of the cord is highly important

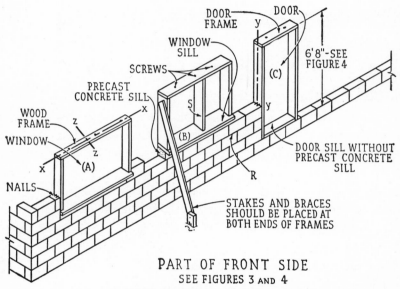

PART OF FRONT SIDE
SEE FIGURES 3 AND 4

FIGURE 15. Placing window and door frames.

because without it the alignment of block cannot be made correctly. Constant checking with the level should be carried on to be sure that each block and the whole wall are exactly plumb. Any deviation from this policy is sure to result in poor construction and the loss of strength and stability.

LAYING UP THE BOND BEAM

As shown in Figure 4, the bond beam is to be laid up in direct contact with the heads of the windows and doors. Until the mortar and concrete in that beam harden and develop strength, it constitutes a load on the frames. In order to avoid any bending of the frames, braces such as shown at S in Figure 15 should be installed. As the temporary braces are wedged into position, the heads of the frames should be checked to make sure they are level.

Figure 16 shows a picturelike section view of that part of the bond beam which is to be laid up near the left-hand corner of the front wall. From Figure 10 it can be seen that the corner block is to contain two bars

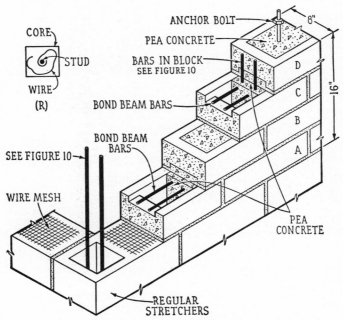

FIGURE 16. Section of typical bond beam.

and that the second block from the corner is also to contain two bars. Those four bars are indicated in Figure 16.

The bond beam is to be composed of four courses of 3 5/8-inch channel blocks as shown at A, B, C, and D in the illustration. Four steel rods are required.

Before starting to lay up the bond-beam block, the cores in the top course of regular stretchers, which do not contain studs, should be plugged, using wire mesh or metal lath. As will be subsequently explained, the plugs will prevent concrete from filling those cores.

All four courses of the beam should be laid up separately, starting with one course leads, and by the use of the guide cord.

Course A. This course is laid with the flat side of the block, as shown in Figure 16, embedded in a mortar joint applied to the surface of the top course of regular stretchers. The joint should be ⅜ inch thick and the same bonding maintained as in other parts of the wall.

When the first course of the beam is complete, two steel rods should be placed, about 2½ inches apart, as shown in Figure 16. At the corners, two of the rods should be bent so that they will extend around the corner at least 24 inches.

Course B. This course should be laid with the open end, or channel, of the block facing downward. As can be seen in the illustration, the first and second courses form a space around the two bars, which can be filled with concrete at a somewhat later time.

Course C. This course should be laid so that the flat sides of the block are down. Then, two more steel rods should be placed as previously explained.

Course D. This course should be laid so that the flat sides of the blocks are up.

Once the fourth course of the beam block has been laid, the beam is ready for concrete.

Anchor Bolts. Figure 6 specifies that anchor bolts shall be placed in the bond beam at 4' 0" intervals. Such bolts can be held in proper position, pending the placing of the concrete in the beam, in the manner explained in Chapter 6, Concrete Foundations.

PLACING STUDS

The long steel studs can be dropped into the proper cores through the cores in the bond beam. They should be of such a length that they will touch the foundation and extend to the top of the bond beam. They can be kept in approximately the center of the cores by means of wire wound around them and extended to the sides of the cores, as shown at R in Figure 16.

PLACING BOND-BEAM CONCRETE

The concrete should be mixed using pea gravel or very coarse sand. If pea gravel is used, the mix can be one part cement, two parts sand, and

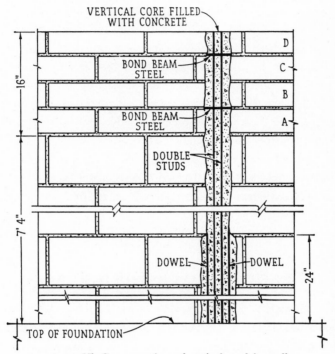

FIGURE 17. Cutaway view of typical stud in wall.

three to four parts gravel. If coarse sand is used, the mix can be one part cement and three parts sand. The pea gravel is recommended. Contrary to the usual recommendations, enough water should be used so that the mixture will *flow* easily. The cores are not large, and, with the steel in them, they offer some resistance to being filled unless the concrete does flow easily.

Figure 17 shows an elevation view of the wall, including the four courses which constitute the bond beam. A cutaway section of one of the vertical studs, from the top of the foundation to the top of the bond beam, shows the horizontal bond-beam rods, the long studs rods, the dowels, and the concrete. This illustration shows how the vertical cores, when filled with concrete and steel, actually constitute columns or studs which have the great tensile strength of reinforced concrete. It can be realized how such columns or studs, with the added help of the reinforced bond beam, tie the whole wall together to produce rigidity and strength which can resist wind pressure, earthquake shocks, and contraction and expansion. There are many such columns or studs in the front wall, as shown in Figure 10. They are tied to the foundation by the overlapping dowels.

The concrete should be placed starting at one corner of the wall. A long steel rod can be used to spade it, so that it completely fills each vertical core and all the spaces between blocks in the bond beam. It is absolutely necessary that the concrete fills all cores and that it completely surrounds all steel rods. If it is placed slowly, with careful spading, the required strength will be assured.

All the required concrete should be placed in one continuous operation, and the walls should not be disturbed for at least 1 week. This length of time is usually sufficient for the concrete to harden and to gain great strength.

BLOCK ABOVE BOND BEAM

Figure 4 shows that the front wall is to extend 16 inches above the top of the bond beam. However, the rear wall is complete with the beam. The difference in wall height is necessary as a means of providing pitch (slope) for the shed-type roof.

Two courses of regular stretchers should be laid up on the top of the bond beam along the front wall. However, the side walls, as shown in Figure 5, cannot be completed using blocks because blocks of diminishing dimensions (or sloping) are not manufactured. Thus between the front and rear walls the side walls must be completed, using placed concrete.

The X part of Figure 18 shows a partial elevation view of the side of the building indicated in Figure 5. The two courses of regular stretchers are illustrated at a and b. The area to be filled with concrete is also shown.

The necessary forms can be constructed as shown in the *Y* part of Figure 18. One 2 by 10 and one 2 by 8, held together with cleats, may be used on each side of the wall. Such forms can be supported by braces

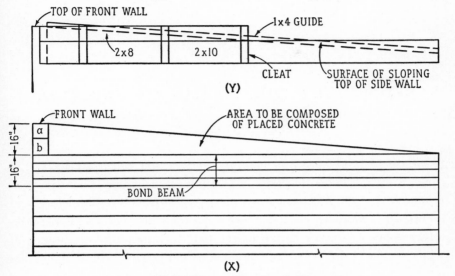

FIGURE 18. Building wall under sloping roof.

extending to the ground or by nailing them directly to the blocks which they contact.

Two long pieces of 1 by 4 may be used as guides and the concrete placed up to their lower edges.

CHAPTER FIFTEEN

Miscellaneous Masonry Projects

There are several aspects of masonry construction which are such common, everyday items that they are all too often slighted so far as good planning and careful construction are concerned. As a result, defects and dangerous conditions exist. For example, both driveways and sidewalks often break and crack to the extent that they mar the appearance of otherwise beautiful lawns and become actual hazards. Stairs sometimes contain one riser or tread which is shorter or longer than the others, and it becomes a stumbling block. Flagstone walks become uneven and out of line and develop edges which cause stumbling. There are, unfortunately, any number of examples all of which point to carelessness in the planning and construction stages of such items.

The purpose of this chapter is to describe good planning methods and proper construction procedures for several miscellaneous masonry projects.

CONCRETE DRIVEWAYS

The planning and construction of commonly encountered kinds of concrete driveways is not difficult if several important points are kept in mind and carefully followed. The following explanations and suggestions are especially presented for inexperienced masons:

Kinds of Driveways. Figure 1 shows four kinds of driveways which are employed in connection with most houses and other small buildings:

320

FIGURE 1 (*W*). This kind of a driveway is the most expensive to construct because of the thickened curbs, the necessary formwork, and the additional concrete required. The surface is convex so that water will flow to the curbs and thence to a prearranged disposal point.

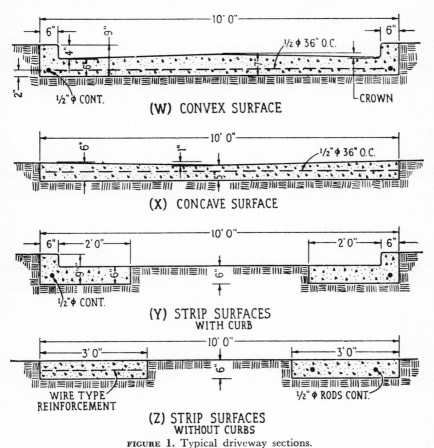

FIGURE 1. Typical driveway sections.

FIGURE 1 (*X*). The concave driveway without curbs is a popular kind because of construction ease and because less concrete is required. The concave surface disposes of water but not as efficiently as the convex kind with curbs.

FIGURE 1 (*Y*). This driveway, while it has curbs, introduces some economy because of the two strips. Many people feel that this kind of a drive is more decorative because of the center section which can be part of the lawn. However, the quick disposal of water is more of a problem which may cause trouble in regions of the country where alternate thawing and freezing occur.

FIGURE 1 (Z). Where exceptionally long driveways are required, this kind, composed of two strip surfaces without curbs, is the most economical in terms of formwork and concrete. However, unless great construction care is used, the strips can quickly assume different levels, crack, and otherwise become objectionable.

Planning. As previously mentioned, and as is the case with all masonry construction, planning is of great importance from the appearance, usefulness, and safety standpoints. Driveways cannot possibly serve all such purposes unless properly planned.

SIZE. First of all, driveways must be wide enough to accommodate an automobile or truck without too much care being required on the part of the driver to keep the wheels on the driveway. This is important as a convenience and because, during wet weather or at times when snow and ice are present, even one wheel off the driveway could cause troublesome ruts in lawns and even mechanical difficulties. Because of this, even at the expense of a few more yards of concrete, driveway widths, where space allows, should be at least 10' 0''. Recommended dimensions for all kinds of driveways are shown in Figure 1. It should be noted that the average thickness recommended is not less than 6 inches.

REINFORCEMENT. Any driveway in any section of the country will be better able to withstand heaving and possible earthquake stresses if it is economically reinforced. This is especially true in regions where soft clay subsoils are present or where long periods of rain or freezing weather must be taken into consideration. Unless provided with expensive footings, the concrete of driveways is not supported in such a manner as to prevent the possibilities just mentioned. The use of a few ½-inch round steel rods or heavy wire is generally advisable. The concrete will then have the tensile strength to resist unusual stresses. In the long run, this constitutes good, safe planning.

EXPANSION JOINTS. Where driveways of any appreciable length are concerned, there are large expanses of surface which absorb a great deal of heat from the sun. This causes some expansion in the concrete. When expansion occurs, contraction is sure to follow. This process sets up stresses which must be relieved in order to avoid the possibility of cracks or buckling. Expansion joints, consisting of creosote-treated wood or other especially prepared materials, are recommended at 30′ 0″ intervals along the length of a driveway.

JOINTS. In order to relieve the monotony of long and unbroken expanses of concrete, shallow joints (made with a joint tool) should be cut into the surface of the concrete at intervals of 6 to 15 feet. Such joints add beauty and often confine damage to limited areas. For example, if a driveway is damaged, repairs can be made between the joints in such a way as to preserve its good appearance. The locations of all types of joints should be marked on the forms before concrete is placed.

SLOPE. All kinds of driveways must have some slope so that water will run off. The runoff can be planned for disposal to street gutters or to sewers. This is especially necessary in regions where rain or thaws are apt to be followed by freezing.

SUBSOIL. Driveways should never be placed on newly filled or otherwise disturbed soils. Such soils are unstable from the all-important support standpoint and cannot be depended upon. It is best to allow at least 1 year's time for newly filled subsoils to settle and compact before driveways are placed on them. The exception to this rule has to do with soils which are composed of sand or gravel.

FILLS. Except where sandy and rocky soils are present, fills composed of sand or gravel are recommended for driveways. Such fills help to dispose of water, prevent possible heaving, and provide the best kind of support.

PARKWAYS

In many cities there are parkways between the street sidewalks and the curbs. The widths of such parkways vary as shown in Figure 2.

For the sake of good appearances, the street curbs should be extended along driveways up to the street sidewalks. Such curbs should be curved to allow easier turning in and out of the driveways.

FIGURE 2 (X). Where wide parkways exist, the curbs should have a

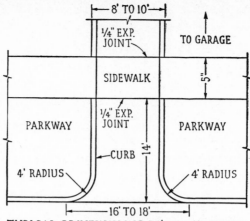

TYPICAL DRIVEWAY FOR 30' ST. 14' PARKWAY
(X)

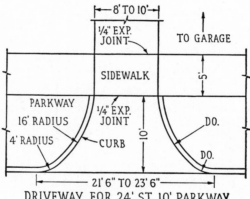

DRIVEWAY FOR 24' ST. 10' PARKWAY
(Y)

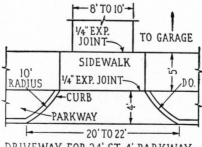

DRIVEWAY FOR 24' ST. 4' PARKWAY
(Z)

FIGURE 2. Driveways in parkways.

short radius so as to conserve lawn space. The width of the street also influences the amount of radius.

FIGURE 2 (*Y*). For medium-width parkways in rather narrow streets, a sweeping radius is recommended so that turning into and out of the driveways can be accomplished with greater ease.

FIGURE 2 (*Z*). For narrow parkways in rather narrow streets, the long radius is also recommended for curbs.

Note that expansion joints (see Figure 2) are recommended in driveways on both sides of the street sidewalks.

FORMWORK FOR DRIVEWAYS

A large percentage of driveway troubles occur because of carelessness in form construction. There has been a great tendency to use makeshift forms as a means of material and labor economy. Actually, and especially for inexperienced masons, the slighting of formwork constitutes one of the worst possible mistakes. It is impossible to construct strong, level, and durable driveways without what may seem like expensive formwork. Experienced masons may be able to form the concrete and properly attend to good appearances. But unless proper forms are employed, the strength and durability of any driveway are bound to suffer. The following formwork suggestions are presented as one means of making sure that driveways have the strength and durability desired:

Forms for Convex Driveways with Curbs. The *X* part of Figure 3 shows recommended form construction for such driveways. The use of 2 by 10s as side forms and 2 by 6s as guides for the template constitutes a sturdy and convenient set of forms. The template can be made from a piece of 2 by 8, the underside of which is formed to suit the contours of the curbs and convex surface. Such a template will slide along the forms and provide an absolutely uniform curb and driveway surface. The *Y* part of the illustration shows details of the template and how it produces the desired contours.

When driveways are to have a 1-inch topping, two templates should be made. One of them can be used to shape concrete 1 inch below the surface levels, and one to shape the topping.

Where curbs are concerned, the concrete should be mixed stiff so that it can be formed with ease. Otherwise, the suggestions made in previous chapters can be followed for the concrete-placing procedures.

Forms for Concave Driveways without Curbs. The suggested construction is similar to the forms for convex driveways with curbs, except

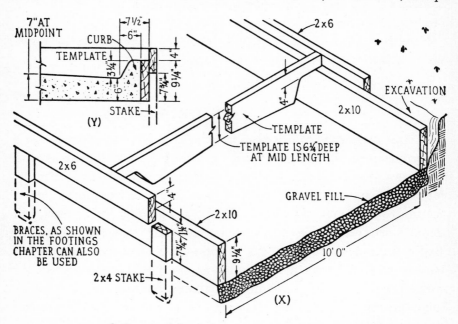

FIGURE 3. Suggested forms for a convex driveway with curbs.

that 2 by 6s are required for the forms and 2 by 10s are used for the guides. A plain template, shaped or sloped toward the center line, is necessary. If no topping is necessary, the template shown in the *Y* part of Figure 4 can be used. When topping is desired, the template with a 1-inch offset, as shown at *Z,* can be used for spreading the concrete base and the one shown at *X* for spreading the topping.

NOTE: In selecting lumber for forms, be sure to allow for differences between actual and nominal dimensions. For example, a 2 by 4 is actually $1\frac{1}{2}$ by $3\frac{1}{2}$; a 2 by 6, $1\frac{1}{2}$ by $5\frac{1}{2}$; a 2 by 8, $1\frac{1}{2}$ by $7\frac{1}{4}$; and a 2 by 10, $1\frac{1}{2}$ by $9\frac{1}{4}$.

Forms for Strip Surfaces with Curbs. The *X* and *Y* parts of Figure 5 show the suggested forms and template for one strip. The other strip can be formed in the same manner and checked for level as explained subsequently.

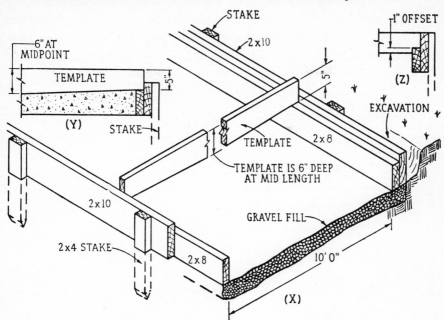

FIGURE 4. Suggested forms for a concave driveway without curbs.

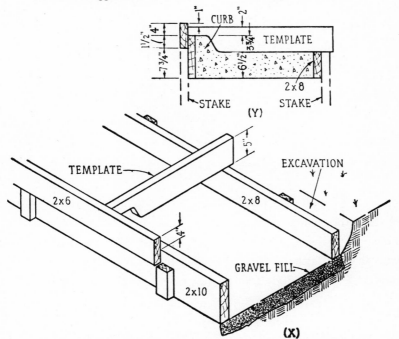

FIGURE 5. Suggested forms for a strip-surface driveway with curbs.

Forms for Two-strip Driveways without Curbs. Figure 6 shows the simple forms for such driveways. If topping is required, one of the strike-off boards should have a 1-inch offset such as shown at Z for use with the concrete base. Remember to allow for differences between actual and nominal dimensions when selecting form boards.

Forms Must Be Plumb. The side forms for driveways must be plumb along the entire length of each and across from one to the other. For

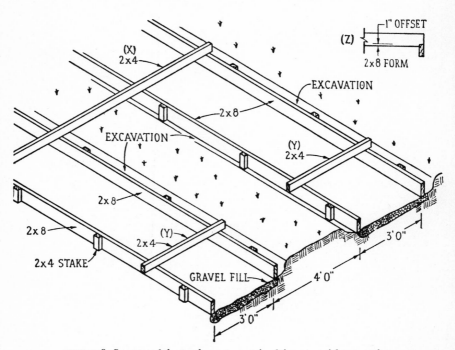

FIGURE 6. Suggested forms for a two-strip driveway without curbs.

forms such as shown in Figure 6, all four side forms must be perfectly level along their entire lengths and all at the same level. Proper level can be achieved by using pieces of 2 by 4, such as shown at *Y,* and a level. Later, all four forms should be checked by using a long 2 by 4, such as shown at *X,* and a level. This process is tedious but important.

The forms shown in Figures 3 and 4 can be checked by placing a level at various locations along the two side forms and then by placing a level on the templates and sliding them along the forms.

DRIVEWAY EXAMPLE

Figure 7 shows the outline of the house indicated in Plates I through VI and typical plans for a driveway and approach sidewalk. All neces-

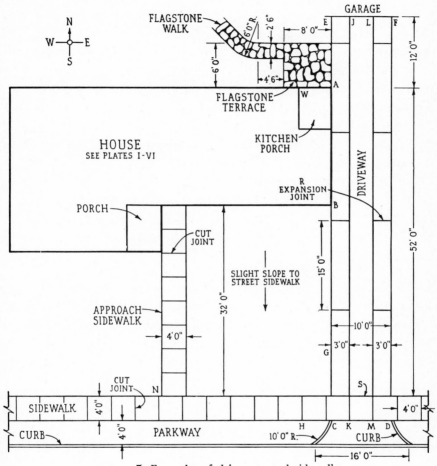

FIGURE 7. Examples of driveways and sidewalks.

sary dimensions are shown. The lot has a gradual and gentle slope toward the street, the parkway has a 6-inch slope, and a strip surface driveway without curbs is desired. It is also assumed that the street sidewalk is in place.

Layout. One side of the driveway is to be in contact with the *AB* side of the house. The layout can be done using that side as a guide.

First, a stake should be driven at point E. The approximate position of this stake can be established by standing at the B corner of the house and sighting along the BA side to the garage. Next establish a stake at the A corner of the house and stretch string, about 12 inches above the ground, from stake A to stake E. To make sure that the angle WAE is 90 degrees, use the right-triangle method explained for Figure 11 in Chapter 5, Concrete Footings.

Next, establish a stake at point C by standing at the A corner of the house and sighting along the AB side to the street sidewalk. Stretch string from stake A to stake C. To make sure that the angle HCA is 90 degrees, use the method of Figure 11 previously mentioned. The stakes at F and D can be accurately located by measuring 10' 0" from the stakes at E and C. Stretch string between stakes E and C and between stakes F and D. These two strings locate the exterior edges of the two strips.

Finally, establish stakes at points J and L and at points K and M which are 3' 0" from the stakes at E and F and from the stakes at C and D. Stretch strings from stake J to stake K and from stake L to stake M. These strings indicate the interior edges of the two strips.

Excavation. The excavation for the two strips, as indicated in Figure 8, should be about 2 feet wider than the strips to allow working convenience as the forms are being constructed. The depth should be equal to the combined thicknesses of the fill and concrete. Care should be taken, by means of plumb lines suspended from the strings, to make the depth exactly the same all along the gradually sloping route of the two strips.

Forms. The formwork shown in Figures 6 and 8 should be constructed next, Carefully measure down from the strings to make sure that all points along the forms are at the required elevation, flush with the surface of the lawn. Care should also be exercised to make sure that the interior surfaces of the forms are exactly under the strings and that the width dimensions of 3' 0" and 10' 0" are maintained. Make frequent plumb checks as previously explained. The tops of the forms should also be flush with the surface of the street sidewalk.

Placing Concrete. Use the proportions recommended in Table 1 of Chapter 4, Concrete, being careful not to add more water than neces-

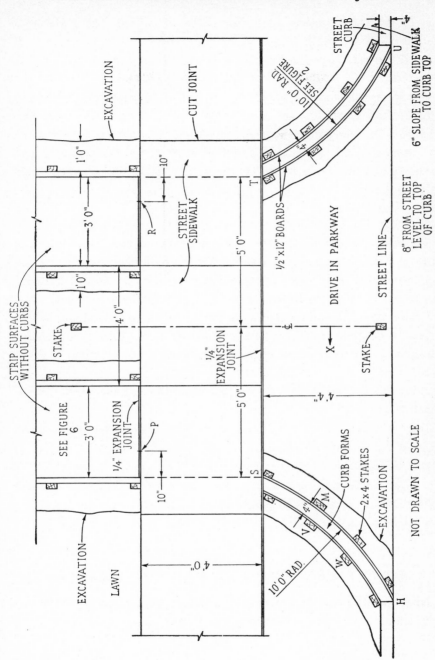

FIGURE 8. Forms for strip surfaces and for parkway curbs.

NOT DRAWN TO SCALE

LAWN

EXCAVATION

SEE FIGURE 6

STRIP SURFACES WITHOUT CURBS

STAKE

EXCAVATION

1' 0"

3' 0"

1' 0"

4' 0"

1' 0"

3' 0"

10"

CUT JOINT

R

STREET SIDEWALK

1/4" EXPANSION JOINT

STAKE

1/4" EXPANSION JOINT

P

10"

S

4' 0"

5' 0"

5' 0"

T

DRIVE IN PARKWAY

STREET LINE

STREET CURB

10' 0" RAD. SEE FIGURE 2

1/2" x 12" BOARDS

4"

U

6" SLOPE FROM SIDEWALK TO CURB TOP

8" FROM STREET LEVEL TO TOP OF CURB

STAKE

X

4' 4"

CURB FORMS

2 x 4 STAKES

EXCAVATION

M

V

4"

W

10' 0" RAD.

H

sary to produce a workable mix. Install the expansion joint material at *R* and *S,* as shown in Figure 7, before starting to place concrete. If colored concrete is desired, it can be mixed as suggested in Chapter 4, Concrete.

Assuming that topping is not required, place concrete in one strip at a time. Start at the garage and work toward the street sidewalk. Add whatever reinforcement is desired. Spade the concrete, especially along the forms, to make sure that it is well compacted. Use a strike-off board, as shown at *Y* in Figure 6, to level the surface. Rake the concrete, as explained in Chapter 7, Concrete Floors, so that fine materials are brought to the surface.

Finishing. As the watery and shiny appearance disappears from the surface of the concrete, use a wood trowel for preliminary smoothing and a joint tool to cut the joints at the places marked on the forms. A 2 by 4 can be placed along the forms and used as a guide for the joint tool. Over the expansion joints, the cuts should extend down to the expansion material. All other joints should be about ⅜ inch deep. Use an edging tool along the forms to make the edges of the strips slightly rounded. Use the steel trowel for final smoothing.

If rough strip surfaces are desired, a stiff brush can be wiped across the concrete to create shallow and parallel furrows.

During warm weather, the forms may be removed after the concrete has had 2 or 3 days in which to harden and gain strength. However, it should be kept moist for several more days. If no hurry is involved, it is wise to wait at least 1 week before removing the forms. By that time, there will be little or no risk of chipping the strip corners and thereby detracting from their good appearance.

DRIVEWAY AND CURBS IN PARKWAY

In order to create a pleasing appearance, the parkway section of the drive should have curbs which match the street curbs.

Layout of Curbs. First, establish a center line as shown in Figure 8. The position of this line can be determined accurately by use of the right-triangle method explained for Figure 11 in Chapter 5, Concrete Footings. Measure 5' 0" on either side of the center line to establish points S and T where the interior curb surfaces will be aligned with the exterior edges of the strips.

The proper radius for curbs in a shallow parkway, as shown at Z in Figure 2, can be found by extending lines from P and R, as shown in Figure 8, to the right and left. These lines are the same as the lawn side of the sidewalk. Drive stakes, which are exactly 10' 0" from P and R, along the extended lines from P and R. Drive a nail in the top of each stake and tie a 10' 0" length of string to each of the nails. The other ends of the strings can be used to establish the arcs SH and TU. This gives the approximate locations for the curb excavations.

Use the 10' 0" strings as a guide in placing stakes for the curb forms. For example, drive stakes at S and M so that the interior face of the stakes will hold the form in proper position, as shown in the illustration.

Locate and place all other stakes for the interior form pieces in the same manner. Once the interior stakes, such as at S and M, have been located, the exterior ones, such as at V and W, can be located by merely measuring the proper 4-inch thicknesses plus the thickness of form material.

Form Material. Boards having a thickness of about ½ inch or less should be used because they will bend easily and stay bent without much strain on the stakes. Once such boards are in place, the stakes can be braced as illustrated in Chapter 5, Concrete Footings.

Curb Slope. The X part of Figure 9 shows the slope from the street sidewalk down to the top of the street curb. The Y part of the same illustration shows the position of the curb forms between the sidewalk and the street curb. These details can be visualized if the reader imagines he is viewing them from the position of X in Figure 8.

Placing Concrete. The forms should be filled with a fairly stiff mix and the concrete well spaded. By means of a little raking, plus the usual spading, the fine material in the concrete, especially if small crushed stone or gravel is used, will come to the top and settle next to the forms.

Finishing. The tops of the curbs should be troweled smooth and the corners slightly rounded. In order to finish the sides of the curbs near the street level where they will be seen, the forms can be removed as soon as the concrete stiffens and such areas carefully troweled.

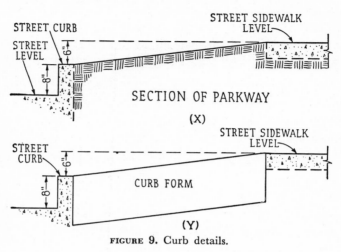

FIGURE 9. Curb details.

PARKWAY DRIVEWAY

Forms are not necessary for placing this section of the driveway. The concrete is to extend from curb to curb and between the edges of the sidewalk and the street. A screed can be stretched from the sidewalk to the street and the concrete spread in that manner. The surface should be raked to bring the fine materials to the surface and one joint cut at about the position of the center line shown in Figure 8. The corners of the concrete, along the sidewalk and street, should be rounded.

NOTE: There are ways by which the curbs and the driveway can be placed in one operation, but the process is apt to provide inferior results and is therefore not included here.

CONCRETE SIDEWALKS

The planning and construction of commonly encountered concrete sidewalks is perhaps one of the easiest of all masonry projects. However, there are certain aspects which should be given careful attention.

Planning. The planning of sidewalks, as with all other masonry projects, should be carried out following several more or less standard rules.

SIZE. First of all, sidewalks should be made wide enough to accommodate the expected traffic on them and, secondly, to have good proportions relative to adjacent lawns and other surroundings. In general, they may be divided into two kinds, namely, street and approach walks. In Figure 7, the street sidewalk runs along and parallel to the street. The approach sidewalk extends from the street sidewalk up to the house.

Street sidewalks are made from 4 to 5 or more feet in width. Approach sidewalks are made from 3 to 4 feet in width. Where wide lawns are

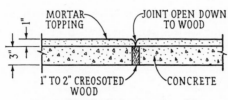

FIGURE 10. Typical expansion joint.

present, a 4-foot width is in good proportions. Narrow walks, under 3′ 0″, never look well in any situation.

The thickness of any sidewalk should be at least 4 inches. This may include a 3-inch-thick concrete base and a 1-inch-thick topping.

CURVES. Wherever changes in direction are required in sidewalks, the curves should be planned to have sweeping rather than short radii. Short curves do not appear graceful, tend to crowd traffic, and make for awkward walking strides. On the other hand, sweeping curves have much better appearance and are easier to walk on.

EXPANSION JOINTS. For the same reason explained for driveways, sidewalks should have expansion joints at about 30′ 0″ intervals (see Figure 10).

JOINTS. For the same reason explained for driveways, sidewalks should have shallow joints at intervals of from 4 to 5 feet. Generally the width of the walks is the joint spacing employed.

SLOPE. Wherever possible, sidewalks should have at least a slight slope so that water can quickly run off. The slope may be from side to side or along their lengths. The amount of slope required, where sidewalks are placed on level ground, need not be enough to notice without the use of a level.

SUBSOIL. The explanation previously given about driveways holds equally as well for sidewalks. In fact, because of narrow widths, this consideration is even more important.

Concrete sidewalks should never be placed on unstable soil where rain could wash out the supporting soil or where settlement is apt to occur. Also, care should be taken not to place sidewalks in such a position that their surfaces are above the surrounding soils. If soil levels are lower than required for sidewalk levels, a fill should be made and allowed to settle and compact for at least 1 year.

If fills are shallow, no more than 1 or 2 feet at the most, they can be compacted by wetting at frequent intervals over a period of a few weeks. Or if the filling soil is sand or gravel, no waiting is necessary.

COLOR. The topping for sidewalks can be colored by the use of mineral oxide pigments. Masonry material dealers can supply coloring of the type which will not fade. See full explanations in Chapter 3, Mortar.

SIDEWALK EXAMPLE

Figure 7 shows a 4′ 0″ approach walk in a lawn which has a natural and slight slope toward the street sidewalk.

Layout. The necessary layout can be made, as explained in the foregoing driveway example, by the use of stakes and string. It is merely necessary to use the right-triangle method of establishing a 90-degree corner at N and make sure that the sidewalk will contact the porch.

Excavation. The excavation, as shown in Figure 11, should be deep enough to accommodate the gravel or sand fill and the concrete. The bottom level of the trench may be checked by suspending a plumb bob from the layout strings and measuring at frequent intervals.

Forms. As shown in Figure 11, the forms can be constructed of 2 by 6s and 2 by 4 stakes. The distance between forms can be constantly checked by means of a strike-off board, as shown in *Y* below. Remember to allow for differences between actual and nominal sizes when selecting form boards.

Both forms should be at exactly the same level along their entire lengths. This can be checked with a strike-off board and a level.

Placing Concrete. Assuming that a 1-inch topping is required, the concrete base should be 3 inches thick. The concrete should be stiff rather

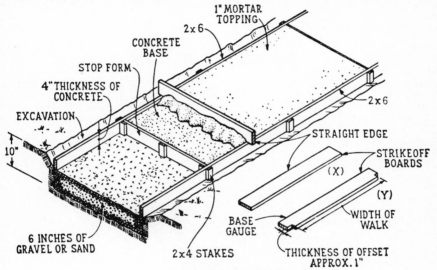

FIGURE 11. Sidewalk details.

than watery. After it has been placed to approximate depth, the strike-off board, shown at *Y,* can be used by sliding it along the forms. The offset will provide a 1-inch space for the topping. The concrete should be well compacted and tamped, if necessary, to complete the compacting. At the end of a day's placing, or where an expansion joint is required, a stop form can be installed.

The topping mixture may consist of one part cement and three parts sand with just enough water to make it plastic and easy to spread. It should be placed before the base concrete has had time to harden.

Dump wheelbarrow loads of the topping on the base and spread it, using a strike-off board such as shown at X in Figure 11. Care should be taken to make sure that the topping is compact against the forms and that no air bubbles are left in it.

Finishing. When the shine has left the surface of the topping, it can be smoothed and finished as previously explained. The corners along the forms should be rounded and the joints made before final set has taken place. Special care should be exercised to keep the concrete moist for several days. This will pay big dividends in the ultimate appearance and durability of the sidewalk.

CONCRETE STEPS

Concrete steps are more complicated than driveways or sidewalks because of the required formwork, the method of placing the concrete, and

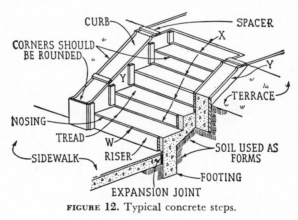

FIGURE 12. Typical concrete steps.

the finishing. It is recommended that inexperienced masons seek the aid of carpenters about the construction and setting of the forms.

Figure 12 shows a combined elevation and section view of typical concrete steps with curbs. The necessary forms are also indicated. Note that a footing is shown and used as a means of better support and to make sure that cracks and settlement cannot occur.

Placing Concrete. The concrete should be stiff so that it can be molded and packed into position quickly. Before starting to place the regular concrete, a stiff mortar mix, such as used for topping, should also be ready for use.

Start placing the concrete at the lower end of the steps so as to fill the footing first. Then fill part of the first step and the curbs on both sides of it. Push the concrete away from the interior face of the riser form and apply a 1-inch-thick layer of the mortar. This will ensure a smooth surface when the forms are removed. Carefully spade the concrete to make sure that it fills all forms completely.

Place concrete in the other steps and in the curbs, following the same procedure. The mortar mix may also be applied to the interior faces of the curb forms as a means of making the sides of the curbs smooth. Allow enough space on the treads and at the tops of the curbs so that topping can be placed before the concrete hardens.

As the topping starts to stiffen, the top corners (nosings) of the steps and the top corners of the curbs should be rounded.

When the topping seems hard to the touch or is not easily disturbed by finger pressure, the forms may be removed.

Finishing. When the riser forms are removed, the corners shown at X may require a little more mortar. Such mortar should be troweled so that the corners are straight and true. The curb surfaces, shown at Y, should be smoothed and more topping added where necessary to make such surfaces straight and true. The tread and riser surfaces should also be smoothed to create an attractive appearance.

FLAGSTONE WALKS

Primarily, flagstone walks are employed as a means of adding decorative effects to lawns and gardens. For such a purpose, where they have little traffic to accommodate, they can be used to good advantage.

Kinds of Flagstone Walks. There are two commonly encountered kinds of such walks. The better, most durable, and more expensive of the two is constructed with placed concrete as a base. The more economical and less durable kind is constructed with sand as a base.

When placed concrete, as shown in the *Y* part of Figure 13, is used as a base, the walks retain all their original beauty and require little, if any, maintenance. When sand, as shown in the *Z* part of the illustration, is used as a base, the walks are apt to lose much of their original beauty, require frequent maintenance, and develop edges which could cause stumbling.

In regions where freezing occurs during winter months, the soil heaving seldom affects walks having placed-concrete bases but raises havoc with walks having sand bases.

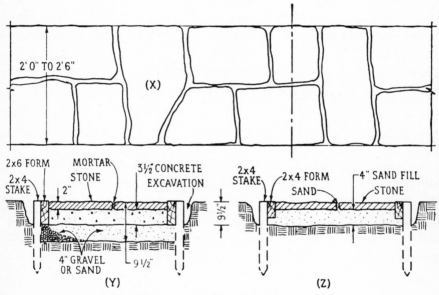

FIGURE 13. Details of flagstone-walk construction.

So far as appearance is concerned, newly constructed walks of both kinds differ only in the shapes and sizes of stones used. Ordinarily, the stones have flat surfaces but may be square, rectangular, or triangular in shape. Walks may be constructed with stones which are all of one general shape, as shown in the *X* part of Figure 13, or they may be planned to include more or less random shapes. The sizes of stones, from the standpoint of area, may vary considerably. Large ones fit the proportions of wide walks, while the small ones are used for narrow walks. Thickness also varies, but 2 inches is common.

Planning. The architectural plans for new houses generally include directions for flagstone walks. Thus masons need be concerned only with construction. However, the following suggestions are presented to serve as a guide in cases where masons are called upon to do both planning and construction:

WIDTH. As previously mentioned, flagstone walks must first serve a decorative purpose. With this in mind, their proper widths depend to a great extent upon the area of surrounding lawns and gardens. Where large areas are concerned, which are over 50 feet wide, the walks may be about 2½ feet wide. For smaller areas, a 2-foot width is in good proportion. Walks having widths of less than 18 inches are not recommended.

ALIGNMENT. Much of the beauty such walks contribute to lawns and gardens stems from the use of graceful curves. Thus, wherever possible, curves should be introduced. Also, wherever possible, the curves should have long radii.

SUBSOIL. Newly filled or disturbed soils, with the exception of sand or gravel, should be allowed to settle and compact naturally before flagstone walks are constructed on them. Otherwise, even walks having placed-concrete bases are apt to settle, crack, and become eyesores instead of the decorative features they were intended to be.

SLOPE. Where any appreciable slopes occur in lawns or gardens, it is recommended that one or two or more steps be constructed so as to keep the flagstone fairly level. The steps may be constructed using thin flagstones instead of topping on the treads.

FILLS. When walks having placed-concrete bases are to be constructed in regions where heavy rains or freezing temperatures are apt to occur, fills are recommended. The purpose of the fills is to dispose of water which might otherwise remain under or near the bases where it could undermine the soil support or cause heaving.

Formwork. The difference between excellent and possibly poor flagstone walks is often accounted for in terms of whether forms were or were not used during the construction process. While it is true that the use of forms adds extra cost to such a walk, it is also true that the extra cost is a good investment. Forms are strongly recommended, especially when inexperienced masons are performing the work.

FLAGSTONE TERRACES

Flagstones may be used to construct decorative and practical paved areas around porches, in patios, around barbecue areas, in play yards, or other places where a form of solid surface is desired.

FLAGSTONE-TERRACE EXAMPLE

Figure 7 indicates that a 6′ 0″ by 8′ 0″ flagstone terrace is to be constructed at the north end of the kitchen porch. (The details of the porch, which is one step above the proposed terrace level, are shown in Chapter 7, Concrete Floors.) Details of the terrace are shown in the *W* part of Figure 14.

Layout. The layout is not difficult because the north side of the house and the west edge of the driveway form two of the terrace boundaries.

First, measure 8′ 0″ from the corner *E* of the house, to establish point *D* along the north side of the house. Then measure 6′ 0″ from the corner *E* to establish point *F* along the west side of the driveway. Point *A* can be established by use of the method explained for Figure 11 in Chapter 5, Concrete Footings. Stakes, batter boards, and string may be used as explained. Or point *A* can be located by finding the intersection of the 6′ 0″ dimension from point *D* and an 8′ 0″ dimension from point *F*. In either case, the forms can be squared up at the corner *A* by means of a carpenter's square.

Excavation. The total depth of the excavation should be 2″ + 3½″ + 4″; or 9½″ to make room for 2-inch-thick flagstones, 3½ inches of concrete, and 4 inches of fill. In other words, the bottom of the excavation should be 9½ inches below the surface of the driveway strip. The bottom of the whole excavation can be checked using a 2 by 4 and a level.

Forms. As shown in the *X* and *Y* parts of Figure 14, the forms can be constructed using 2 by 6s and 2 by 4 stakes. Their top edges must be level all the way around and at the same level as the surface of the drive-way strip. A long 2 by 4 can be placed in the position of *P-P,* and then in the opposite direction, to check finally the over-all form level. In regions where considerable rain or snow is probable, the terrace can be given a slight slope from east to west. This can be accomplished, as the forms are being constructed, by stretching string from stake *F* to a stake

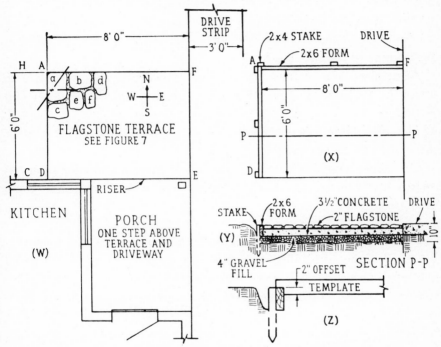

FIGURE 14. Details of flagstone-terrace construction.

driven at the point *H* and from stake *E* to a stake driven at point *C* and then measuring down from them so that the forms, at points *A* and *D*, will be about 1 inch lower than at points *E* and *F*. Use a long 2 by 4 and a level, in the direction of *FA* and *ED*, to check the level.

Gravel Fill. Dump the fill material into the excavation and spread it by means of a shovel and a screed. Place a 2 by 4 across the forms in the position of *P-P* and in two or three other parallel positions, and measure down 5½ inches from its underside to make sure that the surface of the fill is the required distance below the tops of the forms.

Placing Concrete. Use a fairly stiff mix and spread the concrete by means of an offset template of the kind shown in the *Z* part of Figure 14.

Laying Flagstones. This part of the work must be done rapidly and *before* the concrete base hardens. Start by laying stone *a*, as shown in the *W* part of Figure 14. Gently press the stone into position so that it does not wobble. If necessary, in order to achieve a firm bed, pick the stone up and add a little concrete to the bed. The surface of the stone must be at the same level as the tops of the forms. Check this by placing a level in the position of the two dashed lines.

Next, place stone *b* following the same procedure. It is important to check the level of each stone with the form tops and with previously laid stones.

The joints between stones should, as far as possible, be kept to a maximum of 1 or 2 inches wide.

Lay stones along the sides *AF* and *AD* before such interior stones as shown at *e* and *f*. A 2 by 10 plank can be used to kneel on as the stone-laying continues.

Placing Joint Mortar. After all stones have been laid, the joints between them should be filled with topping mortar. It is important that all joints be completely filled. The surface of the joints should, for best drainage and cleaning purposes, be struck flush with the edges of the stones.

FLAGSTONE WALK EXAMPLE

The flagstone walk shown in Figure 7 may be constructed in much the same manner as explained for the terrace in the previous example. It is assumed that the stones are to be supported by a concrete base.

Layout. The position or route of the walk can be staked out and string stretched between stakes to indicate both sides of it. Such layout need not be accurate because part of the beauty of such walks is their irregularity.

Excavation. Assuming that a fill is required, the excavation can be made 9½ inches deep, as shown in the *Y* part of Figure 13. The bottom level of the excavation should correspond to the contours of the lawn or garden surface.

Forms. Forms are recommended as a means of producing a better walk. For straight sections, they can be constructed as previously described for the strips for the driveway, and for curving sections, as described for the curbs. When selecting form boards, remember to allow for the differences between actual and nominal dimensions.

Gravel fill, concrete, and stones are placed as explained in connection with the terrace example. The surface of the walk can be checked by using a 2 by 4 and a level across the forms. Joints are filled as previously explained.

Sand-base Walks. Forms are also recommended in cases where a sand base is desired. Four inches of sand, as shown in the *Z* part of Figure 13, can be placed and checked for level.

The stones should be placed and checked to make sure they are at the proper level and that they do not wobble. Additional sand may be added where necessary to bring stones up to proper level and to prevent wobbling. The joints between stones should be filled with sand.

CONCRETE-BLOCK GARDEN WALLS

Colored or plain concrete block of various sizes and shapes can be used to construct attractive and durable garden walls. Such walls provide privacy and seldom, if ever, require maintenance work or expense.

Kind of Walls. The classification of such walls depends on the sizes and shapes of block used. Figure 28, in Chapter 9, Typical Concrete-block Details, illustrates sizes and shapes of block adaptable to walls of this kind.

Planning. The planning procedures are similar to those used for house walls and foundations, and should be given the same careful consideration.

MODULAR DIMENSIONS. Over-all length, heights, and offsets should be planned so that the size and shape of block used will not have to be cut. This planning is explained in Chapter 8, Concrete Block.

FOOTINGS. Footings are as necessary for block garden walls as they are for house walls and foundations.

JOINTS. All mortar joints should be held to a maximum of ⅜ inch when modular block is involved.

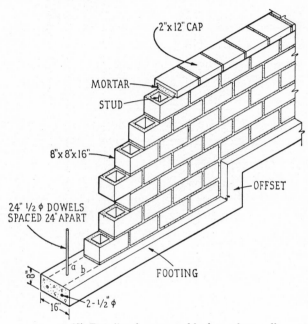

FIGURE 15. Details of concrete-block garden wall.

CONTROL JOINTS. In long walls, as explained in Chapter 8, Concrete Block, at least one control joint should be constructed.

REINFORCEMENT. As with other types of concrete-block walls, reinforcement in the form of dowels and studs is recommended.

CAPS. Some sort of a cap or coping should be used to top off such walls.

Building Garden Walls. Figure 15 shows details of a typical garden wall having one offset due to a slope in the lawn or garden. Such offsets must be planned in multiples of block height so that proper bond can be maintained.

FOOTINGS. The footings, when ordinary-sized block are used, should have dimensions as shown in the illustration. If smaller blocks are to be used, the width and depth of the footings can be reduced according to the W proportions explained in Chapter 5, Concrete Footings. The footing forms are constructed and the concrete placed as explained in the chapter just mentioned.

DOWELS. See Chapter 14, Illustrative Example.

STUDS. See Chapter 14, Illustrative Example.

LAYING UP WALL. See Chapter 10, Concrete-block Walls, Foundations, and Pilasters.

PAINTING CONCRETE-BLOCK WALLS AND FOUNDATIONS

The following explanations, in keeping with the title of this book, include only the use of portland-cement paint:

The demand for white and light-colored block surfaces can be satisfied, to a great extent, by the use of portland-cement paint. When this kind of paint is applied to walls and foundations, it serves two practical purposes: first, it provides attractive finishes, and, second, it helps to make such details more weather-tight.

Such paint is obtained in powdered form in a variety of colors and is mixed with water before application. Class A paint should be used where the surface texture of the block is to be preserved. Class B contains a filler which gives the paint more body and is used to fill open porous surfaces.

The normal exterior wall or foundation treatment consists of two coats of the paint. Where dense blocks are present, which include weather-tight joints, and paint is used only for its color, one coat will generally be ample.

For normal interior surfaces, one coat of the paint is ample. Surface texture and sound absorption values are better preserved when only one coat is used.

Cement paint should be applied to surfaces which are clean and free from oil, oil paint, or any substance which might prevent proper adhesion. Cracks should be cleaned of loose particles, dampened, and filled with a stiff cement mortar.

The surface should be lightly dampened (not soaked) before painting is started so that it will not absorb water needed for proper hardening of the cement paint.

Cement paint should be prepared by mixing with water in the manner and to the consistency recommended in the instructions furnished by the manufacturer. Frequent stirring is necessary to keep the paint powder in proper suspension. A shallow pan 4 to 6 inches deep and 12 inches or more wide provides a good container. When the pan is filled to a depth of only 2 or 3 inches, the paint can be stirred by the painter as he refills his brush.

Brushes with stiff bristles not over 2 inches long should be used to *scrub* the paint into the surface pores of the block. The use of rubber gloves is recommended, since the paint may irritate skin.

The first coat, when scrubbed into the surface, will eliminate any small pinholes through which water might otherwise enter the walls or foundations. Mortar joints should be painted first.

When the first coat has hardened, usually after about 12 hours, the second coat should be applied after first dampening the surface.

Paint should not be applied to frosty surfaces, nor should painting be attempted if the temperature is likely to be below 40°F during the following 12 hours.

When properly applied and cured, cement paint bonds to, and becomes a part of, concrete-block walls or foundations, sealing the joints and block surfaces.

High winds, excessive heat, and sunshine will dry cement paint quickly and render it ineffective as a weather-tight coating unless it is properly moist-cured. The first coat should be kept in a slightly damp condition for at least 12 hours and the finish coat for 48 hours.

Two-tone effects can be produced by applying first and second coats of different but harmonizing colors—as, for example, a first coat of light brown and a second coat of cream. In this treatment the second coat is usually painted lightly over the first, touching only the high spots and being careful not to brush paint into the depressions.

Only light pastel shades of cement paint are recommended for painting exterior block. Other jobs where a particular color effect is desired can be tested by first painting a sample on a separate block or other sur-

FIGURE 16. Applying portland-cement paint.

face. This is important because the color of a painted wall will vary somewhat, depending upon the kind of aggregate in the block, the rate and amount of absorption, and the curing.

CONDENSED SUGGESTIONS RELATIVE TO CEMENT PAINT

The following numbers correspond to the picture numbers shown in Figure 16:

Use pure water and a shallow pan for mixing the paint.

1. Both water and paint should be measured carefully and mixed in accordance with manufacturer's instructions. Use of rubber gloves is recommended to avoid skin irritation.

2. Mix the water and paint until the mixture attains the consistency of thick cream. A simple hand egg beater does the work.

3. Lightly dampen the surface, using a hose having a spray. Do not *soak* the surface. Apply paint while surface is damp.

4. Apply the cement paint with a brush having bristles not over 2 inches long.

5. *Scrub* the cement paint into the mortar joints first in order to ensure complete sealing of joints. Work on the shady side of the building, when possible, to keep paint from drying too fast.

6. *Scrubbing* paint into the rest of the wall finishes the first (seal) coat. Force the paint into the pores of the block; merely spreading the paint leaves pinholes. Allow the seal coat to harden at least 24 hours before applying the finish coat.

7. Cure each coat by keeping it slightly damp with frequent applications of fine spray. Do not soak the surface. Keep moist for at least 24 hours, finish coat 48 hours. Begin spraying as soon as paint has hardened sufficiently to prevent damage.

Index

HOUSE PLANS
Plates I to VI

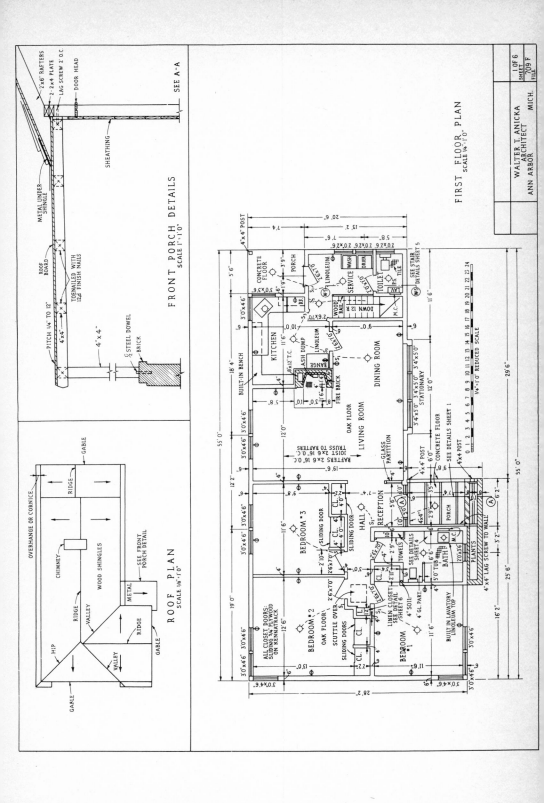

FRONT PORCH DETAILS
SCALE 1"=1'-0"

ROOF PLAN
SCALE 1/8"=1'-0"

FIRST FLOOR PLAN
SCALE 1/4"=1'-0"

WALTER T. ANICKA
ARCHITECT
ANN ARBOR   MICH.

SHEET 1 OF 6
FILE 709 F

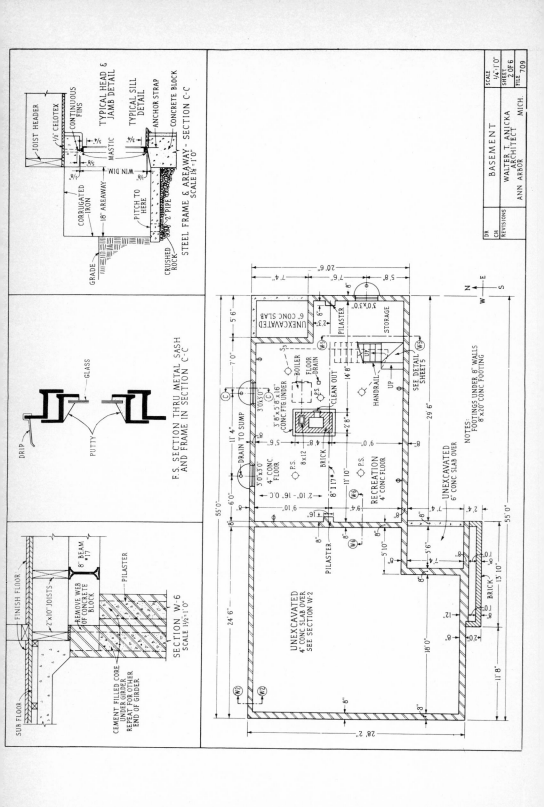

TYPICAL HEAD & JAMB DETAIL

TYPICAL SILL DETAIL

STEEL FRAME & AREAWAY - SECTION C-C
SCALE ¼"=1'-0"

JOIST HEADER

½" CELOTEX
CONTINUOUS FINS
MASTIC
WIN DIM
18" AREAWAY
CORRUGATED IRON
PITCH TO HERE
2" PIPE
CRUSHED ROCK
GRADE
ANCHOR STRAP
CONCRETE BLOCK

F.S. SECTION THRU METAL SASH
AND FRAME IN SECTION C-C

GLASS
PUTTY
DRIP

SECTION W-6
SCALE 1½"=1'-0"

FINISH FLOOR
SUB FLOOR
2"x10" JOISTS
8" BEAM #17
PILASTER
REMOVE WEB OF CONCRETE BLOCK
CEMENT FILLED CORE UNDER GIRDER REPEAT FOR OTHER END OF GIRDER

BASEMENT

WALTER T. ANICKA
ARCHITECT
ANN ARBOR   MICH.

SCALE ¼"=1'-0"
SHEET 2 OF 6
FILE 709

DR
CH
REVISIONS

UNEXCAVATED
6" CONC SLAB

PILASTER
STORAGE
3'0"x3'0"
UP
HANDRAIL
SEE DETAIL SHEET 5
UP
CLEAN OUT
BOILER
FLOOR DRAIN
3'0"x3'0"
3'8"x5'8"x16" CONC FTG UNDER
DRAIN TO SUMP
4" CONC. FLOOR
3'0"x3'0"
BRICK
8x12
RECREATION
4" CONC FLOOR
PILASTER
BRICK

NOTES:
FOOTINGS UNDER 8" WALLS
8"x20" CONC FOOTING

UNEXCAVATED
6" CONC SLAB OVER

UNEXCAVATED
4" CONC SLAB OVER
SEE SECTION W-2

20'-6"
7'-4"
7'-6"
5'-8"
5'-6"
7'-0"
14'-8"
11'-4"
29'-6"
6'-0"
2'-10"-16" O.C.
11'-10"
8'-17"
13'-10"
18'-0"
11'-8"
24'-6"
55'-0"
28'-2"
55'-0"

N
E
W
S

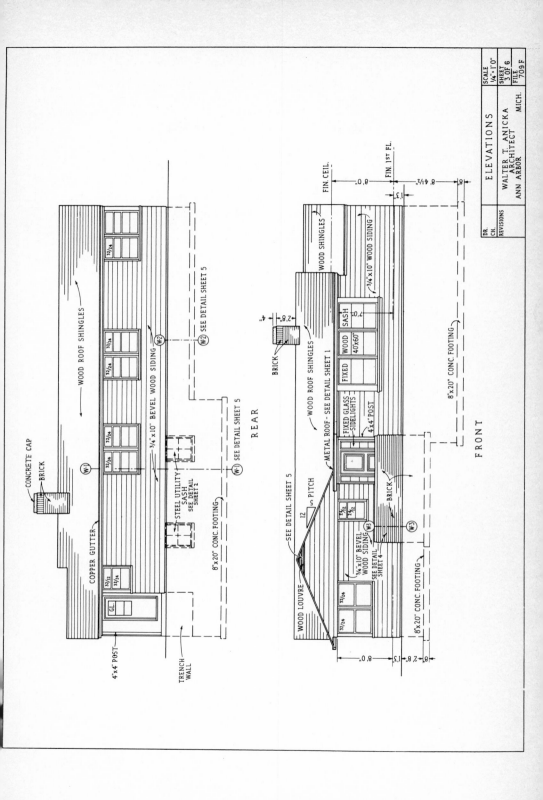

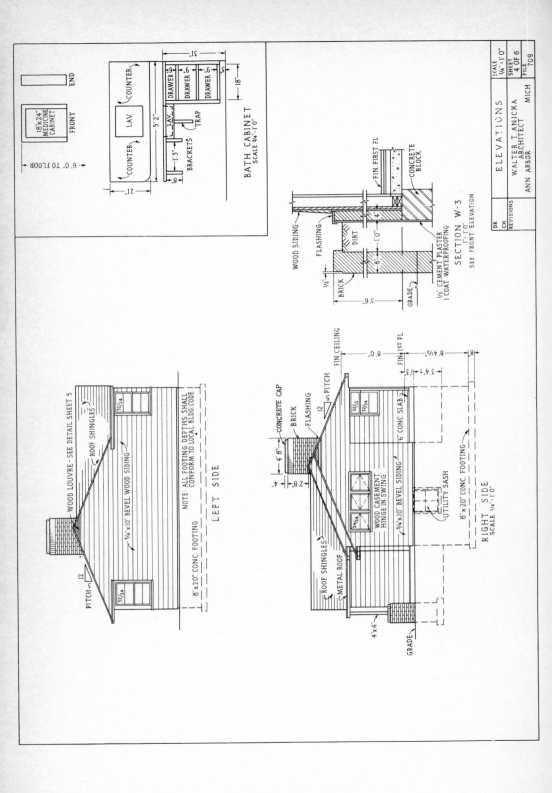

18"x24"
MEDICINE
CABINET

FRONT   END

← 6'-0" TO FLOOR →

COUNTER   COUNTER

COUNTER   LAV   COUNTER

3'-1"

5'-2"

18"

1'-3"   LAV

DRAWER 9"
DRAWER 6"
DRAWER 6"

BRACKETS   TRAP

8"   21"

BATH CABINET
SCALE 3/4"=1'-0"

FIN. FIRST FL
CONCRETE BLOCK
WOOD SIDING
FLASHING
DIRT   4"
1'-0"   8"
1/2"
BRICK
GRADE
3'-6"
1/2" CEMENT PLASTER
1 COAT WATERPROOFING

SECTION W-3
1"=1'-0"
SEE FRONT ELEVATION

WOOD LOUVRE - SEE DETAIL SHEET 5
ROOF SHINGLES
3/2/24
PITCH   12
3/4"x10" BEVEL WOOD SIDING
3/2/24
NOTE: ALL FOOTING DEPTHS SHALL
CONFORM TO LOCAL BLDG. CODE
8"x20" CONC. FOOTING

LEFT SIDE

FIN. CEILING
12 PITCH
CONCRETE CAP
BRICK
FLASHING
4'-8"
4"
2'-8"
3/2/12
3/2/24
WOOD CASEMENT
HINGE IN SWING
3/4"x10" BEVEL SIDING
3/2/24
UTILITY SASH
FIN. 1ST FL
8'-0"
3'-4" ± 1'-3"
8'-4 1/2"
6" CONC. SLAB
8"x20" CONC. FOOTING
8"
ROOF SHINGLES
METAL ROOF
4"x4"
GRADE

RIGHT SIDE
SCALE 1/4"=1'-0"

ELEVATIONS

WALTER T. ANICKA
ARCHITECT
ANN ARBOR   MICH

DR
CH
REVISIONS

SCALE
1/4"=1'-0"
SHEET
4 OF 6
FILE
709

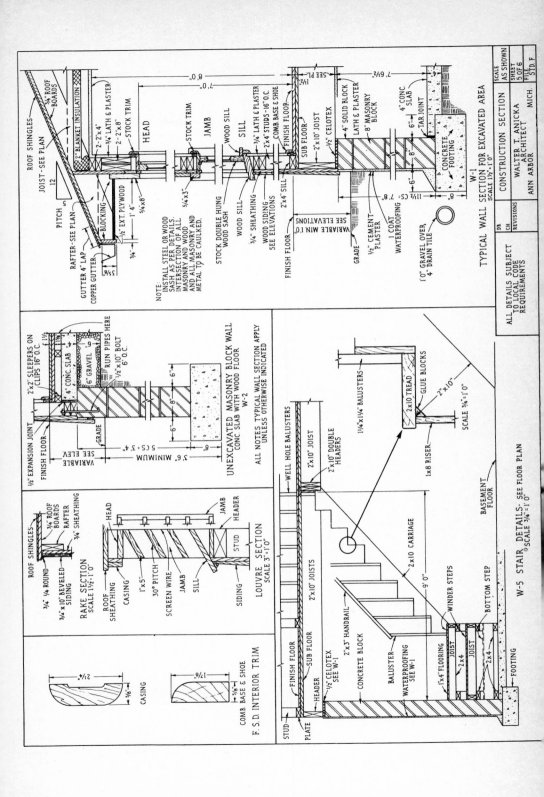

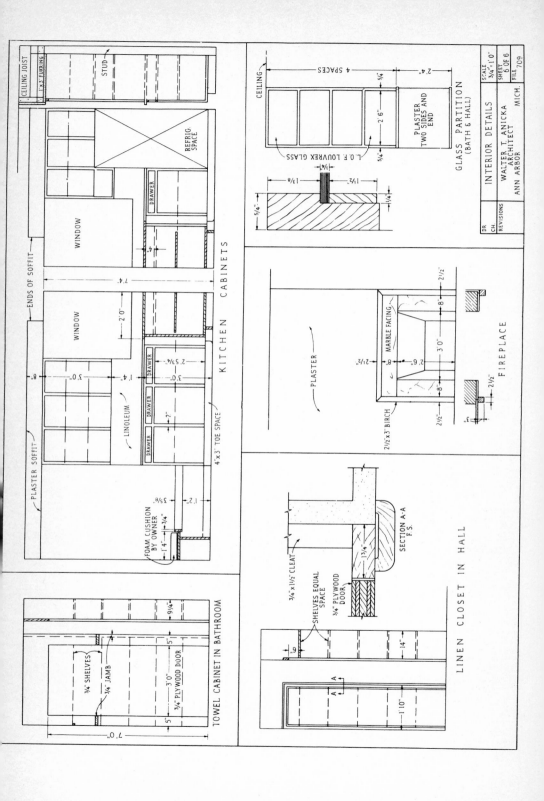

KITCHEN CABINETS

CEILING JOIST
1" x 3" FURRING
STUD
ENDS OF SOFFIT
PLASTER SOFFIT
WINDOW
WINDOW
REFRIG. SPACE
DRAWER
LINOLEUM
7' 4"
2' 0"
8"
3' 0"
1' 4"
2' 5¾"
3' 0"
7"
DRAWER
DRAWER
DRAWER
DRAWER
4" x 3" TOE SPACE
FOAM CUSHION BY OWNER
1' 4"
¾"
3⅝"
1' 2"

TOWEL CABINET IN BATHROOM

¾" SHELVES
¾" JAMB
3' 0"
9¼"
5"
5"
¾" PLYWOOD DOOR
7' 0"

GLASS PARTITION
(BATH & HALL)

CEILING
4 SPACES
2' 6"
5/4"
5/4"
2' 4"
L.O.F. LOUVREX GLASS
PLASTER TWO SIDES AND END
5/4"
1¾"
1½"
1⅝"
¼"

FIREPLACE

PLASTER
MARBLE FACING
2½" x 3" BIRCH
2½"
8"
3' 0"
2½"
2' 6"
8"
8"
2½"
2½"

LINEN CLOSET IN HALL

¾" x 1½" CLEAT
SHELVES EQUAL SPACE
¾" PLYWOOD DOOR
SECTION A-A F.S.
1¾"
A
A
6"
14"
1' 10"

INTERIOR DETAILS
WALTER T. ANICKA
ARCHITECT
ANN ARBOR   MICH.
DR
CH
REVISIONS
SCALE 3/4" = 1' 0"
SHEET 6 OF 6
FILE 709